My own personal onramp to the craft of baking began with my grandmother's biscuits (that's me and her grinding up a squirrel or something below). My decade-long journey to re-create those little knobs of goodness was my first real expression of culinary curiosity and this book is a continuation of that exploration.

The problem is that you can't learn to bake from a book any more than you can learn kung fu from a book. That's because baking requires the transmission of a type of understanding that can only travel the biological bandwidth of personal contact. The reason Americans are, by and large, lousy bakers stems from the fact that while we promote the idea of eating as families, we don't seem too concerned with cooking as families. I, for one, find this odd since eating requires so little skill and cooking requires so much.

So if baking can't be taught by a book, why bother reading a baking book? Well, I didn't say that a book couldn't help you learn to bake. Writing this one has certainly made me a better baker, so maybe—just maybe—reading it will help you become one, too.

I'm Just Here For

MORE FOOD

food x mixing + heat =

baking

Alton Brown

Stewart, Tabori & Chang
New York

Contents

"Things should be made as simple as possible, but not any simpler."

—Albert Einstein

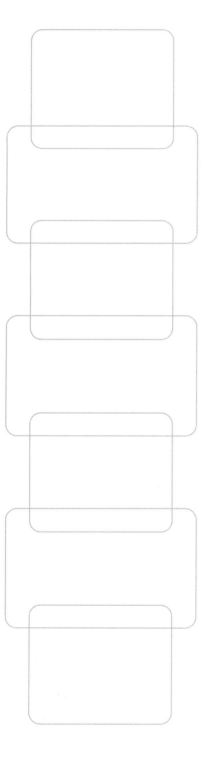

The Devil's Food Is in the Details

Before we get started, I'd like to take a moment to speak of details. Baking is all about sweating the small stuff. Many of the people I speak with who insist that baking is difficult or tedious are actually reflecting upon the "precision" that is necessary to create consistent results. "Cooks" who enjoy facing a pan and tossing in "a bit of this and a bit of that" usually get away with it as long as they don't burn the meat or forget to put water in the rice. Standard everyday cooking is relatively forgiving. Baking is rarely so. In fact, baked goods are a great deal like cars: You can change the wheel covers, put in new mats, and change out the stereo, but if you're going to mess around under the hood, you'd better know what you're doing or you may wind up taking the bus.

In baking, the engine parts are: **flour, eggs, water or milk, sugar, fat, salt, leavening, time, and temperature**. Change from nuts to chocolate chips and your cookie probably won't mind. Change from three eggs to a cup of oil and you're on your own. This whole precision thing also applies to the steps by which you assemble a baked good. How a dough, batter, foam, or custard is mixed together often makes the difference between ending up with a cake or a muffin.

Consider my grandmother's biscuits. For years I tried to clone the tender little jewels of goodness that came out of her oven. At first I assumed that all I had to do was to implement the same hardware and software and follow the same recipe. The resulting biscuits were good, but they weren't quite hers—a fact that everyone in my family made a point of commenting on.

I hypothesized that the secret must lie in her kitchen. I checked elevation data, weather data, and took the temperature in her kitchen, but found no variable that could explain the anomaly. I thought I'd hit pay dirt when I tested her oven (which hadn't been calibrated in my lifetime) and found it ran 50 degrees hot. Alas, adjusting my oven up did no good whatsoever. The biscuits were good, but they just weren't hers.

At this point, I gave up. To heck with biscuits, I've never much liked them anyway. Upon my next visit, I simply slouched at her kitchen table with a cup of coffee, resigned to watch her make her freakishly good and annoyingly unrepeatable biscuits. I watched her remove her rings, slowly twisting them over her arthritic knuckles...a ritual she undertook whenever she thought she might get her hands dirty. Since her hands were always at their stiffest in the morning she rarely made biscuits for breakfast because... hey, wait a minute! The very affliction that caused her so much pain was also the secret to her biscuits. Because she could barely bend her fingers she handled the dough without really kneading it at all. She simply patted it. This is a small detail, yes...but in the end it's the detail that made all the difference in the world.

Details: Don't turn your back on them.

The following pages are pretty much about details—sorting them out and defining them. There are details about ingredients, details about molecules, and lots of details about procedures. But as tedious as all that sounds, I think that—when it comes to baking— details are the truth and I believe I remember someone famous saying that the truth shall set you free.

A Quick Word about Classification

To my mind, the greatest analytical tool in the world is classification. Classifying things leads to enlightenment, and enlightenment to deeper meaning. For instance, I used to make really lousy cheesecake until I realized that cheesecake is not a cake, it is a custard pie. Now I treat cheesecake like a custard pie and everything is fine. There was a time when I did not enjoy Steven Seagal movies. Then a friend pointed out that they are all post-modern Jerry Lewis movies. Now I just can't wait for *Glimmer Man III* to hit DVD. That's classification at work.

In order to understand baked goods I had to figure out a system of classification that not only made sense to me but also brought me to a higher plane of understanding. After doing a lot of reading, a lot of eating, a lot of baking, and a lot of looking at torn up muffins with a magnifying glass, I have come to the conclusion that the best way (for me) to classify baked goods is by mixing method. Not only does this system make sense, it has made me a better baker. So a great deal of this book, including the cute little flaps, are

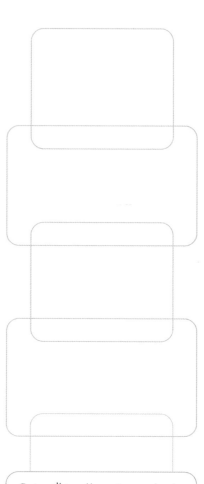

I realize the accepted method of classification— the one used in most cookbooks—is nomenclature-based: pancakes, muffins, rolls, and so on.

I don't think this is any more a "system" than sorting books by color. Names just don't mean that much.

all about grasping the primary mixing methods that make baked goods what they are. Mixing is more important than ingredients and even cooking method (which with baked goods is rather limited). As you read you may find yourself disagreeing with my position that a carrot cake is a muffin or that pie dough is a biscuit, but hopefully you will be stimulated into serious thought about what some might consider trivial.

A Quick Word about the Recipes

The recipes in this book don't look like regular old recipes. They look a little more like the formulas that professional bakers use, except that they don't include bakers' percentages (a method by which all ingredients are expressed as a percentage of the total flour) and they do include volumetric measurements for things that really ought to be weighed. In most cases, the instructions refer back to one of the primary mix methods, which you *will* commit to memory! Until you do however we have these nifty "method" flaps that you can fold into any of the recipes in a given session. They're there to remind you and they're there to save ink and paper. Lots of recipe books basically repeat the same instructions over and over. They do this because it's traditional and because they assume that you are not learning anything. I'm going to assume that you will.

In the assembly and cooking of a baked good, there are many things that are beyond your control. Barometric pressure, for one thing. Here are some other things that you might think are in your control, but they're actually a little out of control: the temperature of your oven, the temperature of your kitchen, the age of your flour, and the hardness of your water. To counteract that which you can't control, you need to control every single thing that you can.

For starters, buy good ingredients and store them properly. Then, measure those ingredients carefully. Take a look at any recipe in this book and you'll notice that most ingredients are represented both by weight and volume. Were it entirely up to me, I wouldn't include volume measurements at all—except perhaps for trace elements like baking soda. But I was advised that no one would buy a baking book without cups and tablespoons, so I gave in. But I want to say right up front that if you're going to bake, you had better get a scale, two if possible. Please. Do it for your food, do it for your loved ones, do it for me.

We'll go over these later, one at a time.

No one. . .not even my mom.

Classifying the Results

A mad scientist attempting to make a marmot from scratch, may (in fact, probably will) slip up and create a horrible mutation that would break out of its cage and wreak havoc on the surrounding countryside. Chased by torch-bearing villagers, the marauding varmint would run home to the scientist, who would attempt to harbor his creation, which would repay him by biting his head off. It's a story as old as science itself.

When attempting to create a muffin from scratch, odds are good that you will from time to time create a horrible mutation. Luckily, lacking a nervous system or even a bad attitude, the odds are good that this aberration will not try to eat your head. In fact, if you close your eyes and take a bite, you may find it to be perfectly serviceable food. I would implore you to not be satisfied. Edible is not good enough.

Classification is an important part of all the sciences, and we (bakers and eaters alike) are obliged to form a set of criteria for each baked good we attempt to produce.

What is a good muffin like? What should sponge cake feel like? Does this crumb structure look like an angel food cake or focaccia?

Balancing the Equation

What this means is that, beyond a correct list of ingredients, a good, sound baking recipe must have instructions written in such a way that—if followed—they will produce the baked good as designed. Said baked good, when constructed as designed, must be what it's supposed to be and taste the way it's supposed taste.

This is not as easy as you may think.

For instance, if I give you a recipe for a chocolate muffin and you follow it exactly and it produces something that you like but wouldn't classify as a muffin, I've failed. If the instructions are written so that, when followed, they produce a lumpy brown mess, I've failed. If you are unable to follow instructions and follow careful procedures in the kitchen, we've both failed. There is, in fact, a lot of potential for failure.

Most of the recipes herein are written as formulas using as few words as possible. I'm not doing this because I'm lazy and don't want to write out every procedure every time. I'm doing it because if you get anything at all out of this book I'm hoping that it will be an understanding that most of the baked goods on Earth follow a mere handful of procedures. Once you see that, you'll start to realize that, just as a man and a chimpanzee have almost 99 percent of their genetic material in common, an angel food cake is more like a soufflé than it isn't. By the same token, once you've got biscuits licked, why not go ahead and apply those same skills to pie crust? They're very similar. The same goes for quiche and cheesecake.

Let me say this right up front: if you're just looking for a collection of baking recipes that you can open to any page and follow at will, keep looking. There are dozens, nay hundreds of tomes packed full of well-tested, reliable recipes that will render tasty results. I'm not saying that the recipes in this book won't or don't work, but you are going to have to actually read the darned thing to be able to use them. Sorry, but that's just the way it is.

If you will read this book, do some thinking, then do some cooking, keep some notes, and then bake a little more, I promise that you will be a better baker. Ultimately I'm hoping that this book can dispense the kind of know-how that will help you to make all your baking better. You'll be more familiar with procedures, have better baking habits, and you might even be able to alter recipes that don't make you happy otherwise.

Occasionally, when we get to a dish that's been especially enlightening to me in my checkered but improving bakery career, we'll take our time talking through things…but I'll try not to get too long-winded.

A Few Quick Words about Good Baking Habits

If baking is about details, you can instantly be a better cook and baker if you just take the time to adopt a few simple detail-oriented habits.

Measure better. This means weighing things like flour and shortening, yogurt, and anything else that cannot be accurately measured in a traditional volume measure. Baking applications are carefully balanced equations and accuracy here matters more than I can tell you.

Aerate your flour and leavening. I rarely add flour to a batter or dough that hasn't been aerated by a food processor, which I use in place of sifting. I never sift. Sifting is for people who don't have processors.

Mise en place. It's French for "everything in its place" and it's a dandy concept. Have all your ingredients measured and ready to go before you start cooking.

Use thermometers. I never bake without an instant-read thermometer. Temperatures matter not only at the end of the cooking but often at the beginning. For instance, the temperature of butter to be creamed, the temperature of eggs going into a batter, and so on.

I like single-subject spirals, but that's just me.

A Few Quick Words on Equipment

Each of the recipes in this book is accompanied by a hardware list of the basic equipment used for the application. Here's a bit more background on what I used in developing the recipes and what I use day-to-day in my own kitchen.

Ovens. I own a GE Profile slide-in range—gas on top and an electric convection oven on the bottom. I don't have a second oven because I don't have room for one. This oven is loyal and trustworthy and friendly.

Please remember that ovens lie. They don't mean to, it's just their nature. They lie because they have to cycle on and off. Over time, they start swinging way off their target temperature. I keep a thermometer in my oven, though I have to say that there aren't many that are easy to keep in position or to read once they're in place. The oven thermometer that I like best is made by OXO Good Grips and features a frosted glass face that allows the light from the oven to pass through for easier reading.

Mixers. Although I keep a hand mixer around, I've come to depend on my electric stand mixer. I've worked and overworked just about everything on the market and have come to the conclusion that the best mixer for me is the KitchenAid Professional 6 (as in 6-quart). It's a good size for the home and has enough torque to…well, let's just say it has enough torque. This is the only stand mixer I have in my home.

Scales and measures. I have two scales, a Soehnle Futura and a Frieling Accu Balance. The latter is very precise and measures to the tenth of a gram, but only handles items up to 8 ounces. That's why I have two scales.

Frieling also makes two of my favorite liquid measures, the tapered Perfect Beaker and the Duo Beaker, which has a mini-measure in the handle—very cool. For sticky or viscous stuff, I use my own push-cups, which are available from my web site (when we have them in stock, that is). To measure really tiny amounts of liquid, one of the syringes that we used to give my daughter medicine when she was a baby works well. They're available at most drug stores.

Oven rack positions

D
C
B
A

THE MIXER ALTERNATIVE

◆ ◆ ◆

If you're an occasional baker, or if most of your baking involves mostly the Muffin and Creaming Methods, the Kitchen Aid Artisan mixer (with the tilt-back head) will take good care of you. But if you bake with great frequency and/or make a lot of yeast doughs that require kneading, then I'd break down and get the Pro 6. Its extra oomph is well worth the money.

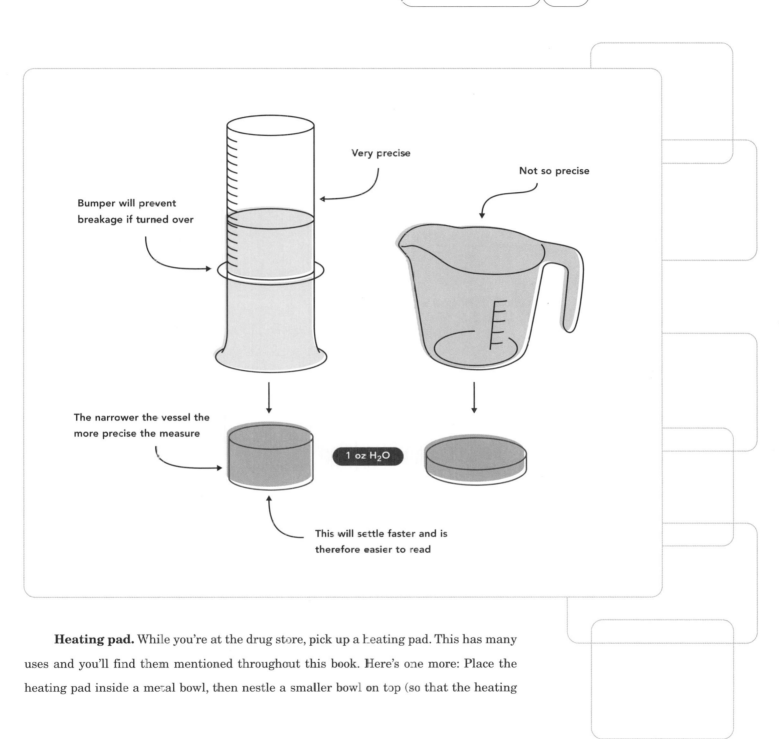

Very precise

Not so precise

Bumper will prevent
breakage if turned over

The narrower the vessel the
more precise the measure

1 oz H$_2$O

This will settle faster and is
therefore easier to read

Heating pad. While you're at the drug store, pick up a heating pad. This has many uses and you'll find them mentioned throughout this book. Here's one more: Place the heating pad inside a metal bowl, then nestle a smaller bowl on top (so that the heating

Weight Versus Volume

CREATING CONSISTENT BAKED GOODS requires consistent measurement. Flour is compressible, as are brown sugar and confectioners' sugar. Salt is equally tricky: a teaspoon of coarse sea salt does not contain as much salt as a teaspoon of kosher salt—which does not contain nearly as much salt as a teaspoon of table salt. It is impossible to measure these ingredients with consistent accuracy by avoir dupois—that is, volume. Heck, I've seen a cup of flour weigh anywhere from 3 to 6 ounces. If you want to measure flour, you have to do so by weight. End of story.

Types of Scales

There are three types of scales: balance, spring, and digital. Forget spring scales; even when they're brand new, their accuracy is iffy. Balance scales are dead-on accurate, but tricky to operate, bulky, expensive, and there's no tare function. Still, their durability and accuracy make them the scale of choice for professional bakers. Me, I'll stick with digital.

My scales easily switch back and forth between standard and metric. Ah, metric—don't be afraid of it. Working with metric doesn't mean you have to convert all your standard American recipes, but if you have a metric scale, you won't have to shun European recipes. This is a good thing, because European baking recipes are generally better than ours because they've been devised under the metric system. And metric is just plain easier: a gram is a gram and a kilogram is a kilogram, and to go from one to the other you just move the decimal three places. There's no dividing or multiplying by 16 and best of all no fractions—none…zip…zilch. For me, this means working in grams is more precise, because I make fewer mistakes.

Since I tend to weigh just about everything I bake with, including liquids, I don't feel as warm and fuzzy about milliliters but I'm trying.

The system of measuring liquids by volumetric ounces.

Meaning the ability to weigh in increments.

Other Reasons Why Scales Are Better

Precision. Meaning your ability to properly interpret what the measuring device is telling you. Even good measuring cups can make this iffy. Scales—especially digital scales—are quite easy to read.

Simplicity. One scale beats eight measuring cups. Modern digital scales have what is called a tare function, which means that after weighing an item (say 8 ounces of flour) you can hit a button, return the display to zero, and weigh another ingredient without removing the first. If all your dry ingredients are supposed to be mixed together, why not weigh them together?

There is a catch. Different digital scales have different capacities for readability; they only display weights down to a certain fraction or decimal point, and that could be a critical problem when weighing small amounts. My rule is that I don't want to attempt to weigh anything on a digital scale that's less than 8 times the readability.

How can you judge readability when you're shopping for a scale? Well, if your retailer goes so far as to put batteries in the display scales, take a few dimes or pennies and just weigh them. That way you can check out the controls and functions. Does it default to metric every time you turn it on (some do); does it easily change between metric and standard? Where do the batteries go and are they a common type? (If I see fancy watch or camera-style batteries in a scale, I steer clear.)

pad is sandwiched between them). Put chopped chocolate in the top bowl and the heating pad to medium (or high, if you have lots of chocolate and don't mind stirring often). The chocolate melts in no time and almost never gets out of temper, which means you now have an easy chocolate coating that you didn't have to jump through a lot of procedural hoops for.

Dry measures. It's Amco Kitchenware all the way for me. They make the best spoons and cups in town—and their spoon rest makes an excellent flour scoop.

Bakeware. My bakeware is mostly aluminum and mostly purchased from restaurant supply houses. My cake pans are made by Allied Metal Spinning and my sheet pans are made by Beacon. Other bakeware in my kitchen is from Frieling, Soehnle, and there's also the stuff that they sell at the grocery store.

Some other tools that make baking easier:

✦ Plastic wrap for covering scales and other things you want to keep clean.

✦ Magnetic timers. I have them stuck everywhere.

✦ Vinyl gloves for kneading sticky stuff.

✦ Cardboard cake rounds from the local bakery.

✦ An oven "fish." Mine is homemade and looks like this:

Push it in

Pull it out

It's just an old piece of broom handle that I whittled on a bit. The front Y is for pushing oven racks in, while the cut out in the back is for pulling them out. It's darned handy.

The Building Theory

It's a well-known fact that a lot of architecture students end up becoming bakers or pastry chefs. Some folks figure it's because they couldn't hack the calculus so they cashed in their dreams of mile-high towers for meter-tall wedding cakes. I like to think that deep down it's because architects and bakers are both structuralists; that is, their crafts are ultimately concerned with form and space.

If you've never thought of your neighborhood baker (if you're lucky enough to have one) as a structuralist, you should: a baked good, be it a muffin, a country loaf, a popover, even a custard, is a lot like a building. Tear open a baked good and you'll see rooms. Sometimes they are very small and uniform, sometimes they're irregular with relatively tough walls, sometimes they are vast and empty. The exteriors that support them are sometimes thick and hard, sometimes light and tender, sometimes springy and moist, sometimes dry...well, you get the point. Architects must understand materials and building methods to create their structures, as should bakers. And of course, bakers have the added challenge of flavor. Now I'm not about to compare a baguette to say, the Gare du Nord, but I humbly suggest that similarities do exist and they are worth considering. Why? Because thinking like an architect will make you a better baker. I know because it made me one.

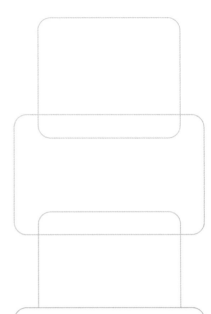

Just so we're clear on this, I'm not a great baker. . .but I am constantly becoming a better baker.

In order to be successful, both architects and bakers must:

✦ Have intimate knowledge of materials and their capabilites.

✦ Have a deep working knowledge of building methods.

✦ Know what it is they want to make.

To my mind, if you want to be a better baker you must:

✦ Have a semi-intimate knowledge of ingredients and their capabilities.

Many bakers bake for years without designing a baked good preferring instead to bake from recipes. Architects often do the same thing.

✦ Have a working knowledge of construction methods.

✦ Know what it is you want to make.

THE STEPS

The way I see it there are five steps to baking anything, and each is pretty darned crucial in its own way. The first three are what I call the "active" steps, during which the baker has the most control. Everything that goes on in baking and cooling is essentially out of your hands.

✦ **Scaling**

✦ **Mixing**

✦ **Make up**

✦ **Baking**

✦ **Cooling**

Scaling

This is the actual measuring out of the ingredients and it's the first step to (I can't believe I'm going to use these words but I am) "baking success." Professional bakers weigh everything from water to flour because they know that even small inconsistencies in amounts of ingredients can result in big inconsistencies of product on down the line. This is especially true of compactible ingredients like flour. I've seen a cup of flour weigh from 3 to 6 ounces, depending on who was doing the scooping. The scooping method is bad! Don't do it, okay? Buy yourself a scale and learn how to use it. Personally, I weigh everything

I can, whenever I can and I use the metric system because recipes written in metric usually don't call for decimals or fractions, both of which I consider to be the work of dark forces.

I usually scale stuff out on a paper plate. I just plop it on the scale, punch the tare function to zero out its weight, then start adding stuff. The plate is great because you can then bend it into a funnel shape to get it into a mixing bowl without mishap.

Mixing

What happens during mixing? First and foremost, stuff gets combined and ingredients like leaveners are evenly distributed, but more importantly, everything comes in contact with water—and water changes everything. Sugar and salt dissolve in it, yeast are activated by it, starches begin to swell or hydrolyze. Also, acids and bases are allowed to mingle, creating CO_2. Depending on the composition of the dough or batter, bubbles will be formed, which will grow into the structures that will define the actual texture of the product. I think that baked goods are best classified by their mixing methods and have spent the majority of this book attempting to prove it.

Make up

This refers to the actual shape that a product will take going into the oven. Does it get poured into a pan or spun into a disk? Is it cut and stacked on a baking sheet or rolled into balls and placed in a muffin tin? Doughs and batters act differently depending on the specifics of their "make up" and bakers should be aware that any changes made to an application (say switching from a 2½ to 3-inch cookie cutter) will make a difference on down the line. In baking, every action has a reaction—though not always one that is equal or opposite. This is not to say that you can't substitute a tube pan for a Bundt pan or a loaf pan for a cake pan, but you do need to know that the results will be different, and not just in shape.

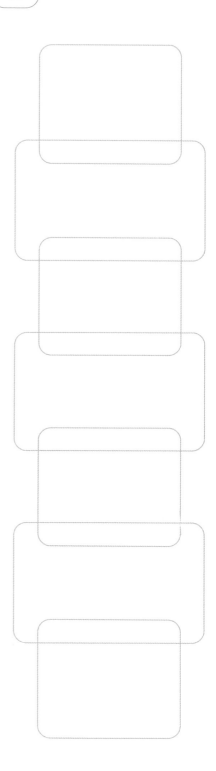

Baking

A lot goes on during baking and almost all of it is out of your control. Yes, you can set the oven and monitor the heat, you can set the rack position and toss in some water for steam, but what actually goes on inside the product is completely out of your hands. What does go on? A lot of things:

+ Gases expand.

+ Fats melt.

+ Starches gelatinize.

+ Proteins coagulate.

+ Yeast and other microorganisms kick the bucket.

+ Water evaporates.

+ Enzymes shut down.

+ Browning takes place.

And by and large, you can't do a darned thing about it.

Cooling

Whenever Aunt Bea caught Opie sneaking a bite of a baked good that was cooling in the window sill, all hell broke loose. That's because Aunt Bea knew that the cooling process was crucial to the quality of the product. During cooling:

+ Bubbles contract.

+ Starches dry and set in a stable structure.

✦ Fats solidify (if solid fats were used in the first place).

✦ Sugars crystallize (at least some of them anyway).

✦ Proteins dry, forming stable structures.

The above can be affected by the baker in some ways by the speed at which the product is turned out to a rack, or left in the pan. Some cakes, like angel food, need to be cooled upside down in the pan so as not to fall under their own weight during cooling. Basically you can think of a baked good as you would a cement sidewalk: it doesn't matter how carefully you mix and form it if you walk across it while it's still wet. ◄

Interestingly enough, concrete generates heat as it "cures" so it could be said to cook itself.

Mixing Is the Killer Ap

Ingredients matter, as do proper pan prep and oven settings, but to my mind, no factor holds greater sway over the birth of a baked good than mixing.

What exactly is mixing, anyway? So many things are achieved during the mixing phase that they are hard to list. Ingredients are integrated, sure, but bubbles are also produced, gluten is or isn't produced. Chemical leaveners begin to release gas, certain solids begin to dissolve in water (salt and sugar among them), starches and proteins begin to hydrolyze (soak up water). Subtle actions taken by the cook during this time can have profound effects on the resulting product.

This can be a good thing...or this can be a bad thing.

If you have a working knowledge of the major mixing methods, odds are everything's going to work out fine even if the formula's off a little. But if you don't know your Creaming Method from your Straight Dough Method, you are at the mercy of the recipe writer. That's unfortunate, because nine times out of ten the mixing instructions comprise the majority of the words in a recipe and the more words that a recipe writer uses to get you where he or she wants you to go, the more potential there is for things to go awry. Words are slippery things: A novice baker could potentially meet his or her doom simply by misinterpreting a single word like "cream," "beat," "fold," or "whip," which—despite being the key action words in many a recipe—are themselves as vague as the plot of *Mulholland Drive*. It's best to know your mixing methods so that when you get hold of a recipe, you recognize what needs to be done in a hurry and don't get caught up in all those pesky words.

LUCKILY, THERE ARE ONLY SIX MAJOR MIXING METHODS:

- ✦ **The Muffin Method**

- ✦ **The Biscuit Method**

- ✦ **The Creaming Method**

- ✦ **The Straight Dough Method**

- ✦ **The Egg Foam Method**

- ✦ **The Custards**

Although I've yet to find any substantial statistics to back this up, it is my deepest gut feeling that if you were to break down the universe of home-baked goods recipes by mix method this is what you'd find:

- ✦ 75 percent of baked goods produced in the home result from only two mixing methods: The Creaming Method and The Muffin Method.

- ✦ Of the remaining 25 percent, I'd say The Biscuit Method and The Straight Dough Method comprise 10 percent apiece. All the other methods fall into the remaining 5 percent.

There are also a few minor methods, but we'll deal with those separately.

Building a Better Baked Good

For me, baked goods are split into two distinct classes: those that are leavened and contain flour, and those that aren't and don't.

CONTAIN FLOUR/LEAVENING	NO FLOUR/NO LEAVENING
Cakes	Custards
Quick Breads	
Yeast Breads	
Cookies/Bars	
Popovers	

This division may seem a bit one-sided until you consider the fact that custards encompass an amazingly long list of goods from curds to puddings to quiches to cheesecakes. But the real differences here are structural. Drop by your local bakeshop and buy a big round rustic loaf. Cut it in half and take a look. What do you see? I see a building with exterior and interior walls that define open space, i.e., rooms. Depending on the building's purpose and the size of the interior spaces, the inside walls may be quite thick or thin, tough or soft. The rooms may be uniform or various sizes. As is true of a building, the materials that go into flour-based, leavened baked goods are chosen for their physical and aesthetic qualities. Understanding the nature of these materials, what they do and how they interact will go a long way toward making you a better baker, not to mention a darned fine architect.

Unleavened baked goods with no flour in them are tougher to find analogies for because they're solid...or so they seem.

Unlike a building whose rooms are defined by the placement of the walls, in a baked good the placement of the walls is defined by the room, or more exactly the "rooms" in a baked good are inflated, then at the right time, the wall material sets so that when the inflating gases are gone the walls remain in place. Whether these walls are thin or thick, tough or tender, whether the rooms they enclose are large or small, whether the flavor is

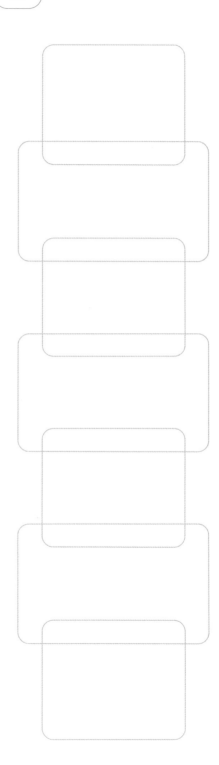

sweet or savory…all these are issues that make various baked goods the unique snowflakes that they are.

I like to think of the ingredients in baked goods as building materials, and to be honest, there aren't that many of them. Let's look at them molecularly, then as actual ingredients.

Flavor aside, everything that goes into a baked good either

- ✦ Strengthens (toughens) it.

- ✦ Weakens (tenderizes) it.

- ✦ Moistens it.

- ✦ Dries it.

- ✦ Leavens it.

Molecularly (but simply) put, here are the major players in baked goods:

- ✦ Protein (strengthens).

- ✦ Starch (strengthens).

- ✦ Fat (weakens/moistens).

- ✦ Water (moistens).

- ✦ Salt.

- ✦ Sugar (weakens).

- ✦ Milk (moistens).

Another way of saying this would be:

Everything that goes into a leavened, flour-based baked good either toughens, tenderizes, moistens, dries, or leavens it.

Proteins and starch, contained in eggs and flour, give a baked good its structural integrity and most of its mass.

Tenderizers like shortening, butter, and the cocoa butter in both chocolate and cocoa powder, weaken the overall structure by lubricating proteins and starches, thus weakening the bonds that make them strong. Sugar tenderizes by holding on to water, and chemical leavenings tenderize by changing the pH of the batter and by increasing the size of the bubbles in the batter.

Moisteners include water, milk, syrups, and eggs.

Driers include powdered milk, flours, cocoa powder, and other starches.

Chemical leaveners—yeast, water, and physically incorporated air—provide the gaseous lift that determines the size and shape of all those rooms.

Although ingredients such as extracts, salt, citrus zests, spices, and liqueurs bring a lot of flavor to the party, they play little or no role in the structural equation of the cake.

Keeping tougheners, weakeners, moisteners, driers, leaveners, and flavorers in balance is what the science of baking is all about. And balance is key. One of the reasons that bake shops employ formulas rather than recipes is that they have to be very precise. Unlike, say, a pot roast, changing any single ingredient in a baking formula will have ramifications in other areas. In baking for every action there may be multiple reactions. If you know something about these reactions, you can take advantage of them.

The Parts Department

As mentioned in the previous section, the list of ingredients that make baked goods "go" is relatively short. In fact, a moderately skilled baker with a good imagination could run a respectable bake shop with nothing but flour, eggs, sugar, butter, shortening, milk, salt, baking soda, baking powder, and yeast. Some chocolate, nuts, fresh and dried fruit, extracts, and the like would certainly be welcome adjuncts, but they wouldn't be absolutely necessary to the process.

The Molecular Pantry

Before we take a look at these relatively familiar ingredients, we should take a moment to look at the ingredients that go into the ingredients—especially the big players in the molecular pantry. Just as the world and everything in it was once thought to be composed of fire, air, earth and water, all baked goods are made up of protein, fat, carbohydrates, water, and air. Then there are "influencers," which affect how these giants function and interact: acids, bases, enzymes, yeast, and salt, among others.

PROTEIN

What it is: A spring made of amino acids.

What it does: Provide structure, strength, and mouth feel.

What it tastes like: Chicken.

Where it is: Wheat flour, eggs (especially the whites), dairy products, meat, soy products, fish, and gelatin.

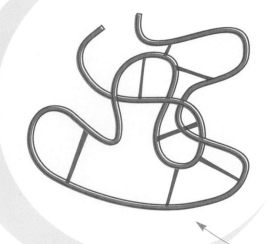

Proteins are long, complex molecules composed of various combinations of amino acids which are…well, let's just call them the building blocks of life as we know it. Since they can be composed of any combination of about twenty aminos, there are a lot of different types of protein.

Remember phone cords—the curly ones that used to connect a phone to the handset? If you still have one, go unplug it and wad it up. That's what a protein molecule looks like in its native or "raw" state. It's held in a wad by a variety of molecular bonds that run along its length.

Now, some of these bonds are weak enough to be broken by heat, acids, salt, or agitation. When this happens, the protein springs open and looks something like this:

The Velcro-like bits are still there, but once the spring is sprung, it's unlikely that the molecule will rebond to itself. Still, proteins rarely travel alone, and if one has popped, odds are a few zillion others have, too. If heat or agitation was involved, these molecules will thrash around and, eventually, bond will find bond and the proteins will coagulate—that is, form a firm, heat-stable three-dimensional mesh. This mesh can take many forms depending on the kinds of proteins involved. Sometimes, different kinds of proteins get together to create secondary structures. Gluten, the elastic yet plastic mesh that makes risen breads possible is a prime example.

CARBOHYDRATES

What they are: Energy storage for plants.

What they do: Provide mass, thicken liquids, and tenderize.

What they taste like: From nearly nothing (flour) to very sweet (sugar).

Where they are: Simple carbs are sugars and syrups of all kinds, and milk; complex carbs are in cereals like wheat.

Cool fact: Despite their diversity, most of the molecules are made up of the exact same building blocks.

CARBS AND YOU

◆   ◆   ◆

Nutritionists and dieticians and doctors and reporters and just about everybody is talking (again) about simple versus complex carbs. What's the difference? Think of it this way:

SIMPLE CARBOHYDRATES— such as sugar, honey, maple syrup and the like—work in the body like cash. They move into our blood system quickly after we eat them and are immediately available to fuel our various tasks. The problem is, if we put simple carbs in our systems but don't use them right away, the organs and glands that regulate our blood have to work very hard to deal with them. This (I'm told) explains the agony of sugar highs and lows, and why so many of us end up with adult-onset diabetes.

COMPLEX CARBS—like those found in legumes and whole grains—are more like a checking account. The body actually has to do some work to liberate these sugars and get them to "clear the bank" of our metabolism. Since the sugars from complex carbs move into our systems more slowly than simple carbs, we don't have to deal with all those highs and lows. Complex carbs are considered "better" because they usually bring other nutrients to the party, like vitamins, minerals, and fiber. Simple carbs are often alone so their calories are said to be "empty."

A Quick Lesson in Carbohydrates

Okay, plants make energy via photosynthesis and in order to store that energy they have to turn it into sugars, little molecules composed of oxygen, carbon, and hydrogen. Think of these as batteries. Now just as there are D cells, C cells, AA cells and 9 volts, there are different types of sugars. We'll say that the simple sugar glucose, the most common plant sugar, is a D cell battery; that fructose or "fruit sugar" is a C cell, and that galactose, which comes from milk and sugar beets, is an AA cell. These are all monosaccharides, meaning they have only one molecule.

To make a disaccharide, or double sugar, we start mixing and matching. If we pair a glucose and galactose together we have lactose. Combine glucose and fructose and we have sucrose, the disaccharide we all know and love as sugar. Maltose is made when two glucoses get together. All of these are simple carbohydrates or "sugars" and they all taste sweet, though they don't all taste equally sweet. In different combinations, these are the molecules that make honey, maple syrup, corn syrup, molasses and malt sugar sweet. Other compounds give them their individual flavors. Although all of these have their place in the baker's pantry, table, or granulated, sugar is by far the most common.

If we line up a few thousand of our D cell glucose molecules like this…

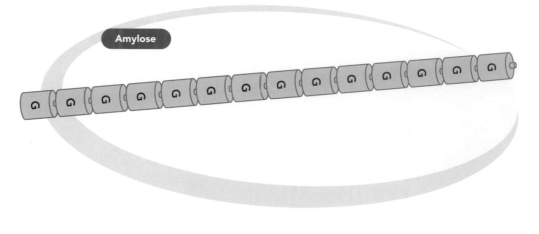

…you'd have yourself a polysaccharide called amylose.

If you break your glucose chain into short segments and rearrange it into a tree-like, branched structure, we'd have the polysaccharide amylopectin.

Mix amylose and amylopectin together in different amounts you have the various forms of starch that make up the bulk of most baked goods.

What Starch Does in Baked Goods

In its raw state, starch is stored in tight little botanical hand grenades that are relatively impenetrable and non-soluble in water—that is until the water gets up to 140°F. Then they begin to gelatinize, that is, the granules begin to soften and soak up water. This water is held along the length of the starches themselves, especially the long, straight amylose. Eventually many of the starches simply burst, spilling their payload like so many overinflated confetti-filled balloons. This process is called gelatinization.

In baked goods, starch serves three purposes:

Starch contributes mass by combining with water and other ingredients. You could think of starch as the concrete of the food world. Like concrete, starch needs to set or dry, which is what the cooling step is all about.

Starch traps and holds water. This includes the water that's squeezed out of proteins during coagulation. One of the tricks often employed in egg-custard–based applications like cream pie filling and cheesecake is that cornstarch is added to the recipe as a precaution against "weeping." Why cornstarch? Unlike flour, cornstarch is almost pure starch, no protein or any other substances to worry about.

Starch provides yeast food. The problem is, those starch granules don't generally break open until they reach 140°F, at which point any yeast present are dead as doornails. Luckily, up to 30 percent of the starch granules in wheat flour are broken during milling, making their energy available to the tiny fungi.

Neither of these structures tastes sweet to us because our taste buds just don't recognize them.

Amylopectin

The word "starch" comes from the Old English stercan, meaning "to stiffen."

FATS

What they are: Energy storage for animals and certain plants, like olives.

What they do: Tenderize baked goods by lubricating starches and proteins.

What they taste like: From nearly nothing (vegetable oil) to very flavorful (cold-pressed olive oil, butter).

Where they are: Butter, shortening, oils, lard.

Fact: Fats that are liquid at room temperature are called "oils." Fats that are solid at room temperature (butter, tallow, lard, shortening) are called "fats."

Fact: Fats and oils are lighter (less dense) than water and they are hydrophobic so they don't mix with water.

Fact: Fat burns in the human body the way its cousin, petroleum, burns in a car engine.

In baking, the word "fat" refers to butter, lard, shortening, and oil. Since it's so vague and general, science types generally avoid the term unless they're describing a colleague who could stand to drop a few pounds. When they want to talk about "fats," they usually speak of fatty acids and the triglycerides they form.

Fatty acids are essentially long carbon chains. Imagine a tram at Disneyland.

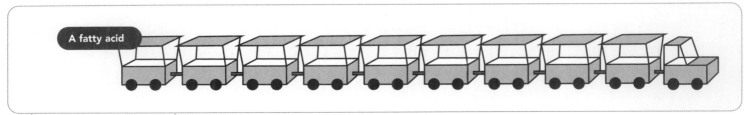

A fatty acid

Just as the tram has seats, these carbon chains have seats, or rather, bonding spaces that can take passengers in the form of hydrogen atoms. If the seats are all full, the fatty acid is saturated. If there is only one seat open, it's monounsaturated. If there are multiple seats available, it's polyunsaturated.

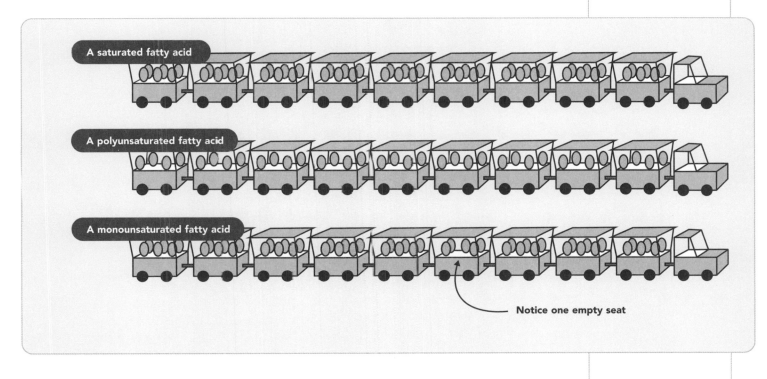

If the train has a kink in it, changing its straight line to an L, then we have a trans fatty acid.

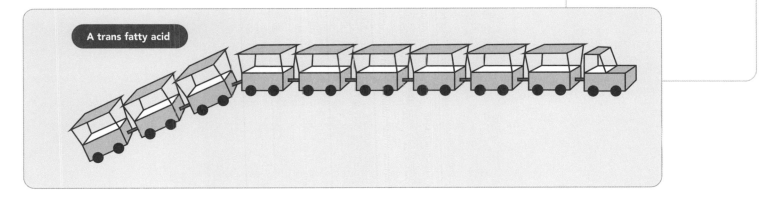

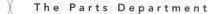

If three of these trains were to pull up alongside each other and hook up with a glycerin molecule we'd have a triglyceride and that's the form that almost all culinary fats take.

> To the best of my knowledge, all culinary fats are triglycerides, but I'm going to cover my bets just in case. . . .

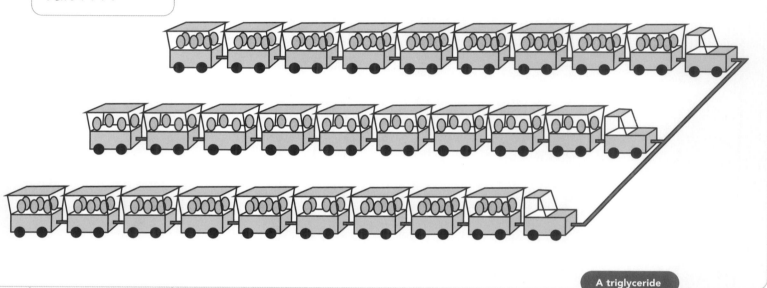

A triglyceride

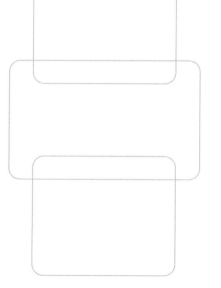

When we talk about a triglyceride being saturated, polyunsaturated, or monounsaturated we're really talking about which one predominates the structure. So in most cases a "saturated triglyceride" is only mostly saturated. The same goes for poly- and mono-unsaturates. Why bother digesting all this information? Because different types of triglycerides perform differently in the kitchen, especially when worked into a batter or a dough.

Triglycerides that are high in saturated fats tend to come from animals and are solid at their intended storage temperatures. They manifest themselves most often in: butter, lard, tallow (beef fat), shmaltz (chicken fat), and, depending on whom you ask, shortening.

Triglycerides that are typically high in unsaturated fats usually come from plant sources and are liquid at their intended storage and use temperature. They manifest themselves in vegetable oils including peanut oil, safflower oil, soybean oil, corn oil, and safflower oil.

What Fats...Sorry, Triglycerides Do in Baked Goods

Fats weaken structure. The best way to understand what fats do in baked goods is to eat a baked good that doesn't contain any fat. A French baguette, for example, is traditionally made with a "lean" or fat-free dough, which is why it's so darned chewy and why it dries out and gets hard so fast. Add some fat and it lubricates the structural elements (the protein and starch) keeping the overall structure tender. Therefore, because fats tenderize they weaken the structural integrity of baked goods.

Fats moisten. Although moisture can only come from water or a water-type liquid, moistness is really just the *sensation* of wetness. Fats that are liquid at serving temperature are very good at creating this sensation—unlike water, they cannot evaporate, nor are they absorbed by starch or proteins, nor hoarded by sugar or salt.

Fats brown. You've seen French fries? Fats can do the same thing to the inside or the outside of baked goods.

Fats elevate heat. Oil heats up fast and it doesn't boil, so it can help to move heat rapidly into a baked good, helping the structure to set quickly.

WATER

What it is: An amorphous arrangement of loosely bound hydrogen and oxygen.

What it does: Water is a solvent and dissolves salt, sugar, and leavening; it hydrates starches and proteins; provides a stable environment for yeast;

The exception to this rule is palm oil, which despite being nearly 100-percent saturated, remains a liquid at "room" temperature. This characteristic makes it very popular with food manufacturers because it is very stable and can therefore keep baked goods moist and tender while they wait on the shelf for you to purchase them. Unfortunately, next to cigarette butts, there aren't too many things worse for you to eat.

Fats throw a big fat wrench in my building/baking analogy because builders usually don't want to weaken their structures.

it carries heat; it is, in essence, the stage upon which the baked good is set.

What it tastes like: Pure water has no flavor or aroma, but since it's such a good solvent it's usually carrying around a bunch of minerals and salts, which do. Some municipal water in the United States tastes and smells of chlorine, which is used to purify it.

Cool fact: The only ingredients that don't include water are pure oils.

Water is in every recipe…every single one. Sometimes it sneaks in without us really noticing it. Milk and its many guises—sour cream, cream cheese, yogurt, buttermilk, cream, and even butter—all contain water. So do eggs. Even flour contains traces of the stuff. The only ingredients I can think of that really don't contain any water at all are oils, which just hate the stuff. Few of the applications in this book actually call for the addition of water, the exception being the yeast-leavened breads, all of which call for water. That's because there are a lot of dry goods to hydrate (proteins, yeast, and starches) and plenty of things to dissolve (salt, sugar). Most importantly, water is also a key player in the production of gluten, the rather miraculous elastic that allows bread to rise.

A Few Things to Consider about Water

Chemicals. If you receive your water via municipal supply, it's been treated with chemicals, chief among them, chlorine. Since it's put there to kill little-bitty critters, it stands to reason that this is not the best place for yeast to be. So consider baking yeast-leavened goods with either bottled mineral water or water that's been run through a filter. Brita and Pūr–brand pitcher/filters do a good job of removing chlorine and suspended solids.

Minerals. Water that contains a lot of dissolved minerals (calcium, magnesium, and the like) is referred to as "hard." Hard water is a problem because it causes mineral buildup in everything from showerheads to hot water heaters to kettles to kitchen sinks.

Hard water can be good for bread because it strengthens glutens in yeast doughs, but if it's too hard it can strengthen the dough to the point where it can't be worked. I've had discussions with folks from regions known for having hard water who said that a pizza dough recipe I developed for a certain television show was too hard to knead. I suggested they use bottled or filtered water and have had several emails reporting back that all is well.

Soft water can be a problem too, though. Most commercial water-softening systems inject a fair amount of sodium into the water and that can pose problems for yeast as well as chemical reactions inside doughs and batters. If you suspect that the stuff coming out of your tap is messing with you and your baked goods, try bottled. Just don't use distilled water. It contains almost no minerals and is too squeaky clean to be good for anything culinary—other than making steam.

pH. Although we usually think of water as being relatively neutral or nonacidic, in those parts of North America that suffer with acid rain, tap water can have a lower pH than usual...that is, it can be acidic. This is generally good news for bread making, as yeast like a slightly acidic environment, but some formulas could be thrown off. So if you know your water to be acidic, buy a filter that's rated to deal with the problem, or again, use bottled.

Basically my stock answer for water is, if you have strange problems with baking, try changing to bottled water before you try anything else—all other things being equal, that is.

LAST, BUT NOT LEAST, AIR

What it is: Roughly 79-percent nitrogen, 20-percent oxygen, and 1-percent argon. Air may also contain 0- to 7-percent water vapor, and less than 1-percent carbon dioxide. The composition changes with altitude. Those

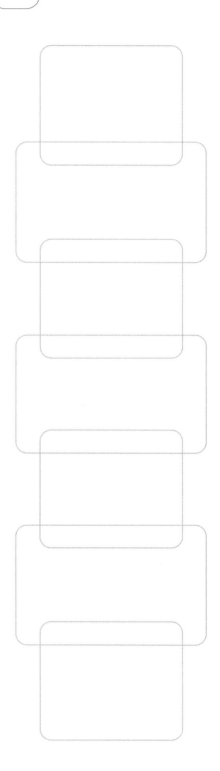

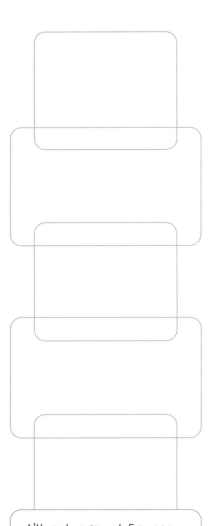

who live in metropolitan areas may also have to deal with various pollutants, which may or may not affect the baking process.

Where we get it: If you're breathing, you've found it.

Fact to consider: The good stuff is getting hard to find.

Its effect on baked goods: Profound.

When air heats, it expands. So if you place a batter or dough that's full of tiny bubbles in a hot oven, guess what? It's going to rise, get bigger, and breathe as the air in those bubbles expands.

Of all the gases in air, oxygen is the most interesting, bakery speaking. That's because O_2 changes just about everything it comes in contact with. Combustion, metabolism, and rust are all made possible by "oxidation." Oxidation is both a good thing and a not so good thing in the cooking world. It's responsible for fats becoming rancid and it's responsible for the maturing of flour, which in turn makes strong gluten formation possible.

Although named for oxygen, oxidation now refers to a wide range of chemical reactions, some of which have absolutely nothing to do with oxygen. Nothing's simple, is it?

The Pantry Pantry

When you go into your kitchen to do a little baking, you obviously don't reach for a bag of "carbohydrate" or "protein." You reach for flour, and eggs, and other ordinary stuff. But many of the staples we work with contain several of the basic building blocks, which is just one of the factors that make baking tricky. Take eggs, for instance: they contain fat, protein, water, and plenty of trace elements that must be considered when developing a baking formula. So, having gone over the basic molecular pantry, let's take a quick look at the basic...well, pantry pantry.

WHEAT FLOUR (LOTS OF STARCH AND SOME PROTEIN)

The Very Least You Should Know:

+ Only wheat flour contains the proteins responsible for gluten formation.

+ Consumer flour styles are rated by their "hardness," or protein content.

+ Commercial flours are also graded by the size of the milled particles and/or what part of the kernel the flour comes from.

+ Flour is porous and will absorb moisture during humid weather. This is why the amount of flour required to construct a particular application is always approximated in the recipe.

+ If you want to accurately measure flour you have to weigh it.

+ There are four main parts of the wheat kernel:
  The germ is the "yolk" of the kernel, and although it is the smallest it contains almost all the fat. That's why wheat germ has to be refrigerated. If it's not, it will go rancid.

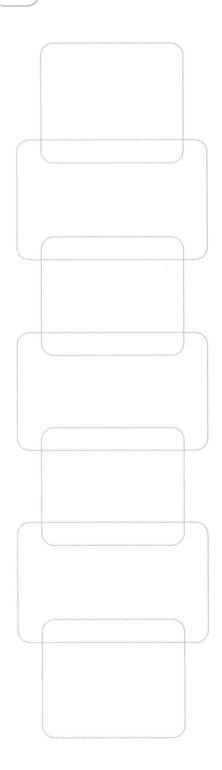

The **endosperm** is the largest part of the kernel and contains all the starch and most of the protein.

The **bran** is a tough coating that is milled into whole-wheat flours. It is insoluble and therefore indigestible…nature's broom, don'tcha know.

The **husk** is composed of several layers of inedible plant matter.

Since it's designed to nourish a baby plant, a wheat kernel contains a considerable armament of protein, but only two are crucial to the baker: gliadin and glutenin. When mixed together with water, these molecules hook together to form **gluten**, a springy mass that is both plastic and elastic. When repeatedly stretched, prodded, and folded (as in kneading) this springy mess straightens out and layers with other glutens to form interlocking sheets that can capture the gases produced by yeast. The closest comparison I can think of is a well-chewed wad of bubble gum. Since gluten is not water-soluble, it makes bread pleasantly chewy. It can also make pie crust, biscuits, and muffins chewy—or downright hard, which isn't a good thing. That's why agitation and/or water content are kept to a minimum in such products.

A Brief Look at the Space Shuttle

A wheat kernel, or berry, is a seed. Underneath several protective husk layers and a coating of bran is the germ and endosperm. Although the germ and endosperm don't actually look like a space shuttle and its fuel tank, it's not an unreasonable analogy given the relative proportions and properties. The fuel tank contains big hunks of protein embedded in starch—everything needed to get the sprout off the ground.

Plastic meaning to bend and stay put. Elastic being the "bounce back to original shape" nature of the waistband on a new pair of BVDs.

Endosperm

Embryo

DNA

The shuttle/germ is not without its own fuel. The germ contains all the seed's fat and all the high-tech stuff like enzymes and other…complicated mechanisms. The proportion of starch to protein in the fuel tank can range from 5 to 15 percent, depending on the type of wheat and when and where it's grown. If the percentage of protein is high, the wheat is called "hard," and the flours milled from it—which are called "strong"—may have a protein content as high as 13 percent. If the protein content is low, the wheat is called "soft," and the flour milled from it "weak." The higher the protein content, the more gluten formation is possible. So, different flours with differing protein contents are marketed with particular baked goods in mind.

Types of Flour

Cake flour is very soft, with as little as 6-percent protein. It's usually milled from the heart of the endosperm, so it has a fine grain. It's also bleached. Bleaching not only lightens the color, it also weakens the proteins so that gluten production is further reduced. I usually don't mess with cake flour unless I'm working with a formula containing a lot of liquid. Bleached flours soak up more water and so produce thicker batter when lots of moisture is involved.

All-purpose is the workhorse flour of most American home cooks. Its protein content usually hovers around 9 or 10 percent, depending on the region. It may be bleached or unbleached, or it may contain a maturing agent like potassium bromate, which I generally avoid because it's a known carcinogen and I like to avoid those whenever possible. Every now and then vitamin C is added to help support yeast growth.

Bread flour is always milled from hard wheat, so it's high in protein (12 to 15 percent) and perfect for yeast breads like baguettes. It comes bleached and unbleached and is occasionally packaged with amylase or malt flour, both of which are enzymes that help to break down starch to make it easier for yeast to consume.

Whole-wheat flour or "graham" flour contains endosperm, germ, and at least part of the bran layer. Although whole-wheat pastry flour milled from soft wheat is available

---

BIGGIE-SIZE CANISTER

◆ ◆ ◆

I keep my all-purpose flour in a fancy flip-top trash can with a slide-out inner container that holds nearly 20 pounds of flour. The lid pops open with the stomp of a pedal and it's nice and wide for easy scooping. Odd as it sounds, few innovations have added more to my enjoyment of baking. No more pesky canisters. Hurray.

Target, $59.99, and worth every dime. . . . . I'm thinking of getting a couple more to store other grains.

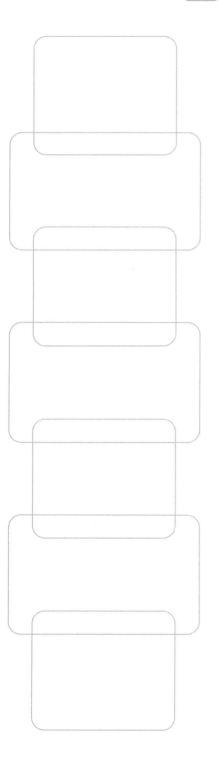

at specialty markets, most whole wheat is milled from hard wheat. Since small but sharp pieces of bran can puncture gluten sheets, many whole-wheat breads contain only 50-percent whole-wheat flour. And because whole-wheat flour does contain fat (from the germ) it can and will go rancid if left out and about, so keep it sealed tight and either refrigerated or frozen.

Durum flour is so darned hard that it's only good for a few things, namely pasta making. It's an ancient breed and despite being loaded with proteins, it's not the kind of protein that makes gluten, so durum is terrible for bread making. It's usually milled into a coarse flour called semolina, which—besides being used in the noodle world—is darned good for sprinkling on your peel before sliding that pizza in the oven. That's because the hard little grains don't dissolve readily and so they don't gum up. Think of them as natural ball bearings.

Specialty flours like oat, rye, soy, potato, and buckwheat can contribute greatly to the flavor and texture of breads—but without wheat flour, there won't be any gluten. No gluten, no elastic bubbles—and that means weak and flabby bread.

Flour is so compressible you could carefully scoop five 1-cup portions and come up with a different weight every time. That's why you should always weigh your flour—always. Besides, weighing is a lot easier than trying to count scoops, especially when the phone rings right in the middle of the procedure.

GELATIN (PROTEIN)

What it is: Dissolved collagen from the connective tissue of animals and from certain forms of seaweed.

What forms it comes in: Sheets and powder.

What it tastes like: If properly refined, nothing. If not properly refined, the animal from whence it came.

What it does: Thickens things like Jell-O and stocks made from animal bones. Unlike starches, which thicken at high temperatures, once dissolved, gelatin must cool to below 50°F in order to set.

The word "gelatin" refers to dissolved collagen found in the skin, bones, and connective tissue of animals. Most of the packaged gelatin available today comes from pig skins, although kosher varieties use other sources.

Gelatin contains eighteen different amino acids joined in sequence to form polypeptide chains called "primary structure." Now, three of these polypeptide chains join in a left-handed spiral creating the secondary structure while in the tertiary structure the spiral winds and folds itself into a right-handed spiral, which results in a rod-shaped molecule known as the protofibril.

Is this information of any use to the baker whatsoever? No. All you really need to know is that gelatin strands are long and thin and that they move around a lot when they're warm. When they drop below 50°F, they slow down and tangle up, resulting in a microscopic mesh capable of holding the liquid of your choice in a firm gel.

What's really cool is that the electrostatic bonds responsible for holding all this magic together are relatively weak. As long as the gelatin isn't very acidic and isn't exposed to heat in excess of 150°F, it can be melted and reset over and over again.

Dried gelatin comes in powdered and sheet form. I like the sheets because the

A *polypeptide* is a single, linear chain of amino acids.

Unless you have access to that newfangled Internet thing.

gelling power is higher, but they're darned hard to find. I usually stick with Knox powdered gelatin, which is available just about anywhere groceries are sold.

Powdered gelatin is reliable and easy to use as long as you play by its rules. Since it's dehydrated, it needs to be soaked in **cold** liquid before it's heated and dissolved. It won't hydrate in hot water so don't go thinking you can speed things up by simply sprinkling it in hot water. The general rule of thumb is to sprinkle the gelatin over ten times its weight in cold liquid and allow it to sit for ten minutes to "bloom" before heating the liquid (the same is true for sheet gelatin). The liquid doesn't have to come to boil in order for the gelatin to dissolve, a gentle simmer will do just fine.

Once the gelatin is dissolved in the blooming liquid, you have to be careful how you introduce it to the liquid you plan to thicken. If you just dump it in, you'll end up with a thin liquid with tiny rubber balls in the bottom. The trick is to slowly whisk some of the target liquid into the gelatin solution to dilute it. Then you whisk that back into the rest of the target liquid. It's just like tempering eggs for a custard (see pages 294–295), but in that case it's usually the target liquid that's hot.

EGGS (FAT, PROTEIN, AND SOME WATER)

The Very Least You Need to Know:

- ✦ Eggs are the most versatile ingredients in the kitchen.

- ✦ An unrefrigerated egg ages a week in a day. Keep eggs refrigerated, in their carton, in the back of the fridge.

- ✦ Egg whites contain mostly water and protein; no fat.

- ✦ Egg yolks contain protein, less water, and all the egg's fat—as well as phospholipids that act as emulsifiers.

- ✦ Eggs are major structure builders.

Occasionally someone will ask if there's any single food that can teach more about cooking than any other. I like being asked this question because I've got a fast, snappy answer: eggs. The seed of *Gallus domesticus* is, to my mind, the most flexible and powerful culinary device on Earth. Eggs never cease to amaze me and, although they are delicious on their own, they really shine when they jam with the rest of the culinary jazz combo that is the wet works. They provide support to flour-based forms, they emulsify batters, and they create the magical mesh that is custard. They do it all.

In fact, there is probably not enough room in this entire book for me to say all that I want to say about eggs, so I will do my best to avoid a mega-rant of Proustian proportions.

The Egg: What Exactly Is It, Anyway?

A chicken egg is designed to provide a growing chicken embryo with everything it needs, from protection to nourishment—and it's a darned good design, too. The only problem (from the egg's point of view) is that a vast majority of chicken eggs are never fertilized. But that doesn't stop chickens from laying them. The average American laying hen (a Single-Comb White Leghorn) starts laying at about twenty weeks of age and cranks out an egg every twenty-five hours for about two years, after which she retires to a nice semi-private coop in Florida with a view of the sea and a tidy pension. ◄

FACT

✦   ✦   ✦

In addition to eggs, how many foods can you think of that come in their own convenient single-serving containers? Bananas, coconuts, and...Ben & Jerry's cartons... that's about it.

If you really want to know what happens to chickens that fall behind in production, see *Chicken Run*. The truth is, most laying chickens on commercial poultry farms are treated like commercial machinery, which I'm not entirely comfortable with. If I ever manage to move to the country, I will raise my own chickens for eggs. Until then, I buy organic.

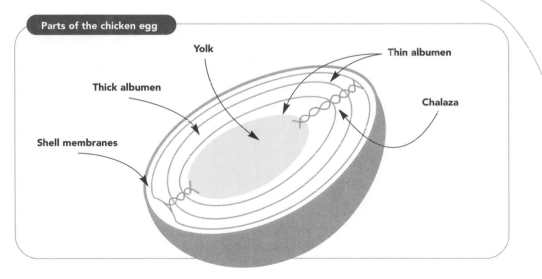

Parts of the chicken egg

Yolk

Thin albumen

Thick albumen

Chalaza

Shell membranes

FOOD FOR THOUGHT

◆ ◆ ◆

The two big power players in baking are wheat kernels (from which flour is milled) and eggs, and although they seem unrelated, they are actually amazingly alike. Both are designed to house an embryo, which is contained, or intended to be contained, within a fatty germ surrounded by nutrients meant to feed the new life within.

Consider...

Ironic...don't ya think?

The carnauba tree is a member of the palm family. It produces a very hard wax that's used in everything from cosmetics to car wax.

Shell. Despite expressions like "walking on eggshells," the truth is that an egg shell is a tough customer. That's partly because it's made up of more than 90 percent calcium carbonate and partly because of design. But not all eggs are created equal. If the hen's diet is high in calcium, phosphorus, and vitamins, she may produce eggs that a grown man cannot crush in one hand—unless he's got arms like Popeye. This is especially impressive when you take into account that the shell is not solid. It's pocked with thousands of pores that allow air in, and moisture and carbon dioxide out. When it's shiny and new, the egg is covered with a natural protective coating called the cuticle, a waxy substance that seals the pores, keeping microbial invaders out. Eventually the cuticle dries away, leaving the egg vulnerable. For this reason many egg packers spray eggs either with mineral oil or carnauba wax.

By the way, the only difference between brown eggs and white eggs is color. White chickens by and large lay white eggs and brown chickens, brown. The color is only on the very surface of the shell. A little fine sand paper can turn a brown egg into a white one lickety-split. Brown eggs do not taste different from white eggs...or if they do, it's got nothing to do with color. It's because of feed. Like people, chickens are what they eat. I happen to think that organic eggs taste better because the hens that lay them are usually fed better chicken chow (or is that chick chow...I forget).

Shell membranes. There are two of them: an inner and an outer, both of which encase the egg and protect it from bacteria. As the egg cools after laying, the insides contract. That creates a vacuum that pulls air in through the porous shell. The bubble that forms between the two membranes is called the air cell.

The air cell is a clever device, which as far as I know is the only part of the egg's interior formed after exiting the hen. The space forms as the inner portion of the egg cools and contracts. Luckily the shell is porous, allowing air to fill what would otherwise be a vacuum that could break the egg. As the egg ages, the egg continues to lose moisture and carbon dioxide, and the air cell grows larger. If you've ever hard-cooked an egg and peeled it only to find a big flat spot (usually on the small end) that's where the air cell was.

The egg white, or albumen, makes up almost two thirds of the egg's liquid weight and is composed of water, protein (lots of it), vitamins, minerals, and some sulfur. The white contains no fat whatsoever. Because its interior is relatively alkaline, eggs are very vulnerable to bacteria.

The white is actually three different structures: two thin layers surrounding a thick layer. This is most noticeable when a fresh egg is broken out on a plate.

The chalaza (ka-lay-zeh) is that thing that looks like a white cord hanging out of the yolk. The word is Greek for "hailstorm," which sheds absolutely no light on anything. The chalaza is formed from egg white that's twisted as the egg chugs down the hen's oviduct. Its sole purpose is to keep the yolk centered in the egg.

The fresher the egg, the more pronounced the chalaza. It's perfectly fine to leave this intact if you're scrambling or poaching eggs, or beating them into a batter. But if you're whipping up a custard, these biological bungee cords will have to be strained out before cooking or you won't achieve a smooth texture. Since the chalaza dissolves as the egg ages, you could simply skip the strainer and wait a few weeks for your eggs to get old. However, as eggs age, the membranes around their yolks become so weak that they're almost impossible to separate.

The yolk contains much less water compared to the white, but it contains all the egg's fat, cholesterol, phospholipids, and the emulsifier lecithin. Although it looks smooth and yellow, the yolk actually consists of a continuous phase of water and protein in which are suspended a whole bunch of yellow granules. Anyone who's ever choked down an overcooked hard-boiled egg has felt these granules on their tongue. These granules are amazingly complex stews of stuff including various types of lipids (the larger molecular family that includes fats).

What's an Egg Weigh?

If you're like me and prone to losing count of how many eggs you've put where, here are the weights used by many pro bakers. So if you don't know whether you've got 12 or 13 egg whites in that bowl, just weigh them, then divide by one of the numbers on page 48 and you'll have your count.

divide by one of the numbers on page 48

DATES, CODES, RANDOM NUMBERS

❖ ❖ ❖

Somewhere on the carton, usually the end, there will be a bunch of numbers and letters, usually in two rows. One of these will be a "use by" date. Ignore it, for it is completely random. The other row will generally start with a letter followed by four small numbers. That's the plant code. Then there will be some numbers in a bigger point size. This is the Julian date of the day the eggs were packed. A Julian calendar runs from 1 to 365 in a non–leap year. So May 10 is 130, August 25 is 237, and December 1 is 335. When you consider the fact that most eggs are packed within a couple days of hatching, these Julian dates are the best way to judge the day-age of the eggs. But as Indiana Jones says, "It ain't the years, it's the mileage"...so if the eggs haven't been stored correctly during that time, they could be a lot older, quality-wise.

EXP OCT 30 C

WEIGHTS		
large egg	50 g	1.75 oz
large egg yolk	20 g	.7 oz
large egg white	30 g	1 oz

I have never, ever, seen a recipe call for a jumbo egg, or a pee wee egg, or any other kind of egg other than large, and that is what you're most likely to find at the megamart. However, if you find yourself attempting to bake a cake when all you've got are pee wees in the house, just separate the whites and the yolks and start weighing.

Grade: Like the grading of beef, the grading of eggs is a voluntary process. Ungraded eggs are by no means less wholesome or nourishing or, well, cookable, it just means the packer has decided not to pay the extra money. If he or she does they're likely

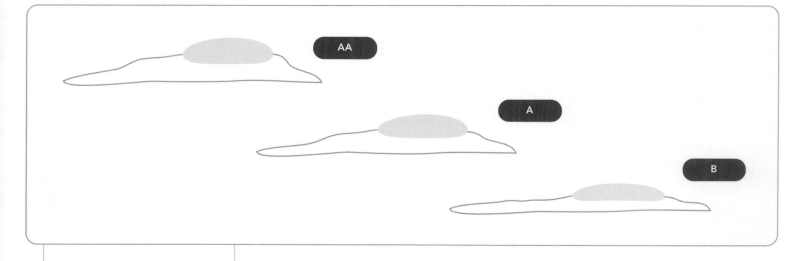

to receive one of the following: Grade AA, Grade A, or Grade B. When candled or examined with a high-intensity backlight, AA eggs display the smallest air cells. When cracked onto a plate or "broken out," a Grade AA egg will have a very pronounced, thick white and a very round (perky?) yolk. The only difference between a AA and an A egg is that the air cell is a little bigger in the A and the thick albumen spreads out just a little.

A Grade B egg is easy to spot because it has a pretty big air cell and the yolk is almost flat. An egg may be graded B for countless reasons, from the type of feed to the breed of chicken, but most of the time it's a factor of age—and by that I don't necessarily mean time.

The rate at which an egg goes downhill has more to do with handling than time. To an egg, a day at room temperature is the same as a week in the refrigerator. The warmer it is, the faster the membranes that separate the different parts of the egg deteriorate. Properly stashed in their carton, in the back of the fridge (those egg slots on the door are evil—they spend too much time wagging out into a hot kitchen) eggs will keep an amazingly long time. AA grade eggs will drop to A eggs in about a week but won't descend to B for about six weeks. After that they're still perfectly edible but I wouldn't do much more than scramble them. Once they're that "old," eggs are all but impossible to separate because the yolk typically breaks the moment the shell is breached.

Speaking of breaching eggs, I'm going to steal from my first book in order to emphasize the following:

> Depending on how fresh they were when you bought them.

How to Crack an Egg

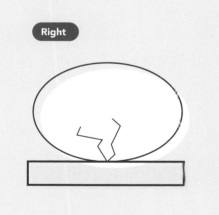

Right

Flat blow cracks egg cleanly.

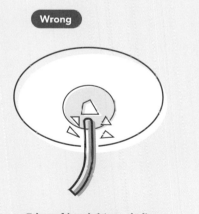

Wrong

Edge of bowl drives shell fragments and any outside germs deep into egg.

FACT

✦ ✦ ✦

If you get a broken piece of shell in your egg, the best tool to remove it is another piece of shell.

> If refrigerated, the average egg will literally dry up before it "goes bad" or rots. I've kept eggs for ninety days and they've been just fine. . .but I'm not saying you have to.

FACT

✦ ✦ ✦

Since it is actually designed to support animal life, the egg contains a nearly perfectly balanced collection of amino acids. Meat doesn't even come close, but then meat isn't designed to be a food, it just ends up that way.

A FEW OTHER THINGS TO KEEP IN MIND ABOUT EGGS:

✦ ✦ ✦

✦ They contain nearly 75 percent moisture, so adding more to a recipe adds more water, too.

✦ The proteins in eggs can be used as an edible glue. Paint it onto a dough with a brush and whatever you stick to it will be there forever once the baking's done.

✦ Eggs enhance color, flavor, and the body, or mouthfeel, of many foods. Although the texture part of the equation is still a bit of a mystery, the flavor intensity probably stems from the fact that fat carries flavor very well to our taste buds.

✦ And don't forget, eggs just plain taste good. In fact, nothing tastes like egg but egg.

What Eggs Do in Baked Goods

Eggs provide structure. Yes, they contain fat, which is antistructure, but that smidgen of destructive power pales in comparison with the egg's ability to build. There are two structural components in eggs. The first is protein. Although they are nowhere near as elastic or plastic as the gluten-forming proteins in wheat flour, egg proteins can create two different types of structures: gels and foams.

Egg Gels. When "cracked" or denatured by agitation, acid, or heat, egg proteins weave together in a three-dimensional net that can capture and hold moisture, starch, fat—whatever.

Without this structure, many baked goods—especially those that contain little or no gluten, like cakes—would lie in the pan, hollering, "I've fallen and I can't get up." The really groovy thing about egg proteins is that the gel they produce can be firm, soft, or just about anywhere in between, depending on how much liquid is added to the equation and how much they're agitated (compare a quiche with a custard sauce).

Egg Foams. The same protein mesh that makes cakes and custards possible can also be used to reinforce tiny bubbles that, when grouped together, become a foam. When this foam goes into the oven, the air inside expands. If the formula is a good one and your technique is sound, the proteins in the bubbles will set the structure before the air is lost, so in this case the eggs are providing both structure and leavening. This phenomenon makes miracles like meringues, soufflés, angel foods, and genoise possible. Although egg-white foams are the most acclaimed, yolk-only and whole-egg foams are also possible and darned useful.

Many cookbooks refer to the tenderizing power of egg yolks. Truth is, egg yolks are not tenderizers at all, it's just that the structures they create are more tender than those formed by whole eggs. So, if you replace a whole egg or two with yolks, you will find that the product is less tough, not more tender.

Beyond Structure. Eggs, especially yolks, contain lipids that are powerful emulsifiers. Lecithin is an especially powerful emulsifier, and egg yolks contain more of it than just about any other natural source. Emulsifiers are the official peace-keeping force in the baked-goods world because they can molecularly marry fats to water-based liquids. Consider a pancake or cake batter that contains a fair amount of fat and either milk or some other liquid. Bringing them together into a cohesive batter would require a huge amount of physical agitation were it not for emulsifiers, and that agitation would be bad because once the liquid is in a flour-based batter, only toughness can come of prolonged beating. Egg yolks and the emulsifiers they contain allow baked goods to come together faster, and unless a lot of kneading is required, faster is better.

SUGAR (STARCH)

The Least You Should Know

+ Table sugar is 99.9 percent sucrose, a disaccharide made up of one fructose and one glucose molecule.

+ Besides sweetening, sugar's primary job in baked goods is to tenderize.

+ Sugar's second job is to leaven through aeration.

Honey, maple syrup, corn syrup, molasses, malt syrup—sugars all. Although they possess individual traits, they all contain high levels of glucose, a simple single sugar, or monosaccharide, that's the most common sugar found in nature. It's also called "blood sugar" because it's most often used to sweeten Tasmanian blood punch, which is drained from…just kidding. It's called "blood sugar" because glucose is the sugar that's in your veins right now…unless you're a vampire in which case…never mind. The other common monosaccharide is fructose, or fruit sugar. When two simple sugars hook up, they can form a *disaccharide,* or double sugar, that we call sucrose or table sugar.

Refined from both sugar beets and sugarcane (the only plants that produce sucrose in harvestable amounts), table sugar is 99.9 percent pure sucrose. Although a lot of other

FACT

The only difference between glucose and fructose is that glucose has a six-sided carbon ring while fructose's ring has five. They have the same chemical formula.

A Very Brief History of Sugar

WHEN DID MAN START TO CRYSTALLIZE SUGAR from the juice of *Saccharum offic-inarum*, sugarcane? No one is sure, but there are records of "honey without bees" being produced along the banks of the Ganges River at least 2,500 years ago. Back then, refining was still in its infancy, and nobody had a clue how to remove the snow-white crystals we all know and love from the dark, brown, gooey molasses or the cane syrup that never fully crystallized. So sugar came in big, sticky, brown loaves or cones that had to be scraped rather than sprinkled.

From India, the knowledge spread west to Persia. There, Arab merchants—who recognized sugar's true potential—tinkered with refinery technology until they managed to produce crystalline sugar. Then they completely monopolized the sugar trade worldwide. Their smartest move: transporting it across the northern rim of Africa and into Spain. Then all they had to do was let it trickle into Europe at about $500 a spoonful. Finally Europe got fed up being stuck in the dark ages while the Arabs were living the good life, so they launched a thinly veiled religious operation called "the Crusades." One of their goals was to get sugar. So they stole into Spain, swiped sugarcane clippings, and took them back to Brussels and London—where they didn't do so well. But eventually the Europeans made

such cones are still available today in ethnic markets and in the Hispanic section of many major megamarts. They are called "pilloncillos."

their way to the New World and the subtropic lands of the Lesser Antilles and the Caribbean—lands where cane could thrive. Thus were born the Colonies.

Cane begat plantations, and plantations required labor—rum, molasses, slaves—the infamous triangle. Eventually sugarcane spread to Cuba and then into southern Florida, where most of our supply comes from today.

Although a German chemist by the name of Andreas Marggraf developed a process for extracting sugar from beets back in the mid-eighteenth century, it would take a war to bring real interest to the project. Among the many problems that Napoleon ran into when he decided to wage war on the planet was fueling all those retreats. Just kidding…the problem was feeding all those soldiers. Sugar was in especially short supply. He needed a domestic supply, and he found it in beets. The beet-sugar industry got a second shot in the arm at the close of another war. America's Civil War had put an end to slavery, which then made the harvesting of cane impossible. Afterward a mechanical means had to be developed. Today the biggest beet-sugar producers are the U.S., the Netherlands, Germany, and France. Sugar beets don't grow in the tropics.

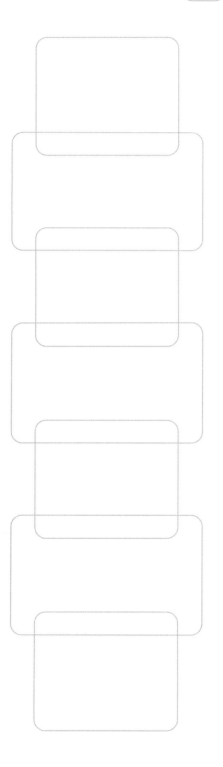

sugars and syrups have uses in the kitchen, this is the sugar we think of when we think of sugar, and it's the one we're going to get into here.

Common Market Forms

Granulated/Fine sugar. Got a canister set on your counter? Go to it and open the middle one. That's granulated sugar. Empty, you say? Go to your pantry. Open it. Look for a 5-pound bag that's not flour. That's it. Yes, it looks a lot like fine salt, but the light refracts off it differently. If in doubt, taste.

Superfine/Castor sugar. Basically, this is granulated sugar only finer. It's called "castor" in the UK because it's fine enough to be dispensed with a castor. (Would it kill them to just say "shaker"?) Since it dissolves quickly, it's especially useful in applications like meringues. Should you buy some? No. You should take granulated sugar for a 30-second spin in your food processor instead.

Powdered sugar/Confectioners' sugar. Also called "icing sugar" in Canada. These sugar crystals have been pulverized to dust, or near dust. The ratings of X, 6X and 10X refer to how fine the dust is, with 10X being the finest commonly available. I keep 6X around because it's fine enough for icings but not so fine that it simply dissolves if dusted onto a finished dessert (or French toast) as a decoration. Powdered sugars are packaged with cornstarch, which absorbs the moisture that would otherwise turn the contents into a white brick.

And speaking of clumping...

Brown sugar used to be made by adding molasses to sugar syrup before crystallization. But today brown sugar is simply white sugar with molasses added after the fact. With a little practice, you can make your own at home by adding ½ tablespoon of molasses to 1 cup of granulated sugar. Brown sugar often clumps because the molasses dries, literally gluing the granules together. The old trick of sticking a piece of apple in a bag of hardened sugar works just fine, but so does a wet cotton ball.

Although it's not very common, I have to mention **turbinado sugar,** because the stuff has a complex flavor I really dig. Instead of being created by an additive process, turbinado sugar is made from unrefined raw sugar. Since the "impurities" removed by refining are locked in the crystal rather than poured onto the outside, turbinado doesn't clump as badly as mass-market brown. A slightly lighter version of turbinado called **demerara** is especially popular for coffee in the UK. Then there's **muscovado** or **Barbados sugar,** which may be a little finer than turbinado, depending on who you ask. Like "gourmet" salts, many of the attributes of artisanal sugars are up for discussion. Although there are supposed to be rules governing how these things are graded, like the pirates' code, they're really more a set of guidelines than rules. Savvy?

There's also a special sugar made in Japan called ***wasanbon toh***. It's made from a very special type of sugarcane called *chikutoh,* which is short and skinny. The fine crystals of this sugar are the product of a long, intensive, hands-on process. From those I've spoken with who have witnessed the process, I don't think it's as hard as making a samurai sword, but it's close.

What Sugar Does in Baked Goods

First and foremost, **sugar provides sweetness.** But flavoring agent is only one of the roles that sugar plays in baking.

Sugar tenderizes baked goods in two different ways. First, it gets in the way. For example, sugar molecules get between egg proteins in a custard, thus slowing the coagulation process. Sugar is highly hygroscopic, or water-loving, so it can also tenderize by grabbing water and preventing it from being used by tough structural components like protein and starch. This same water-hoarding capability helps to stave off retrogradation or "staling" by keeping water and starch together.

Sugar preserves by binding up water so that it's not readily available to the microbial life forms that would love to get hold of it.

Sugar leavens. As we'll soon see in the discussion of the Creaming Method, sugar leavens by punching tiny holes into solid fat. In the heat of the oven, these bubbles expand, thus lifting the product and creating its internal texture. Without crystalline sugar, most cookies and cakes would be impossible to make. Sugar also aids leavening by

slowing the movement of water into starch, and that means batters and doughs containing sugar can expand further before setting.

Sugar stabilizes egg foams by holding onto and dissolving in the water contained in the bubble walls. That's why angel food cakes are so much easier to make than soufflés, which usually don't contain as much sugar—unless, of course, they're dessert soufflés, which are easier to make than savory soufflés.

Sugar browns. When cooked to a high temperature, sugar breaks down into hundreds of smaller compounds including color and aroma compounds. This is called *caramelization* and it tastes, well...like caramel. Sugars also brown via the famed Maillard reaction. In Maillard browning, sugars, more complex carbohydrates, and certain amino acids react together in the presence of heat to create new flavors and colors. Although caramelization and Maillard reactions are often confused, it's important to note that Maillard browning can take place at much lower temperatures than caramelization—and without added sugar. When you get a nice brown sear on the outside of a steak, that's Maillard at work, not caramelization.

SOLID AND LIQUID FATS (FAT...AND SOMETIMES WATER)

Butter

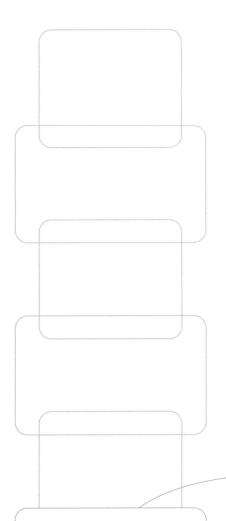

Technically, butter can be made from the milk of any mammal, but only cows and their kin (yaks, for instance) produce enough fat for the butter-making process to be practical.

Butter is made from the cream...usually from a cow. In its liquid state, cream consists of tiny globs of butterfat floating in milk. When agitated (or churned), these fat globules begin to stick together. Eventually you end up with a big lump of fat surrounded by liquid. Drain off that liquid, knead the fat glob smooth, and you've got butter, which—being mostly saturated fat—is solid up to about 90 to 95°F.

But butter isn't all fat. Most of the butter produced in the United States and Canada is only 80 to 85 percent butterfat. The other 15 to 20 percent is water, which can be good and bad news. During baking, some of this water will turn to steam and can be used to leaven. The downside is that before it turns to steam, it's just plain water and accounted for in the overall formula. If not, there could be trouble because water activates starches and proteins and could therefore contribute to strengthening rather than tenderizing. What's left behind are milk solids (powerful browning and flavor agents) and emulsifiers.

Despite its high saturation, butter can go rancid quickly because of its water content. Salt is often added to butter in part as seasoning but mostly as a preservative. Unsalted or "sweet cream" butter is usually wrapped in a foil paper to prevent air contact, whereas salted butter usually comes wrapped only in paper.

"Sweet cream" butter is a misnomer. It doesn't contain any sugar or other sweeteners—it just isn't salty.

Three things butter does very well:

1. Butter adds unique flavor.

2. With a melting point just below the body temperature of a (normal) human, butter melts smoothly and slowly in the mouth, thus creating a rich mouthfeel that can't be matched by any other fat.

3. The milk solids in butter will brown via the Maillard reaction.

However, the truth is that when it comes to baked goods, butter really isn't very efficient and it's kind of a pain to deal with, because:

✦ Butter isn't 100 percent fat, so when using in a recipe you always have to consider the water content and what trouble it might cause.

✦ Butter has a very narrow window of plasticity and is best worked with between 65° and 70°F. Any colder and it's hard as a rock; by the time it hits 96.8°F, it's completely liquid.

✦ Butter goes rancid quickly. If you want to keep a big supply on hand, keep most of it tightly wrapped and frozen.

Skip the wrapping and your butter will quickly taste just like the freezer.

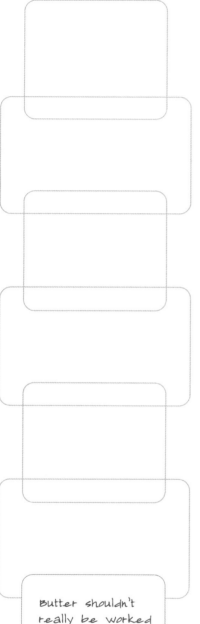

Butter shouldn't really be worked until it hits 65°F, but I want some advance warning.

Working with Butter

The biggest problem you'll face with butter is getting it from a stick to the shape you need without it turning into goo in your hands. Here are some techniques that I use.

Cutting butter. Before I do any cutting of butter, I open up the wrapper and stash the butter in the freezer for a half hour. If you don't unwrap it, the foil or paper will freeze into the butter on the ends, which is very inconvenient.

If I need to cut butter into flour for pie crust or biscuits, I use the large holes on a box grater to just grate it into the flour. Just spray the face of the grater with a little non-stick vegetable spray beforehand so that it doesn't bind. A vegetable peeler works pretty well, too. If you're really serious about keeping the butter solid for as long as possible, chill the bowl and flour along with the butter.

Melting butter. Who wants to break out a pan just to melt some butter? If I'm mixing a small amount of butter with other ingredients, say eggs and milk for pancakes, I just put the butter in a metal mixing bowl and put it over low heat on the cook top. Heavy-duty stainless steel bowls will handle the heat just fine.

A microwave can melt an entire stick straight from the refrigerator in 60 seconds or less. The problem is that the water inside the butter can heat so quickly that it turns to steam. This means it suddenly needs to occupy about a thousand times more space than it previously called home, so the butter blows up all over the oven. For reasons beyond what my meager mind can fathom, wrapping the butter in a single layer of plastic wrap (I'm a StretchTite fan, but that's just me) seems to prevent this and, by holding heat inside, next to the bar of butter, it actually seems to accelerate the melting process. When the dinger dings, just reach in—with tongs, please—and extract the plastic. Very neat.

Creaming butter. When I know I'm going to cream butter, I remove the sticks from the fridge a good half hour beforehand and shove a probe thermometer into one of them. My Polder has an alarm that can be set to go off at a particular temperature, so I just set it for 60°F and wait.

Shortening

The least you need to know:

- ✦ Shortening remains plastic at a much wider temperature range than any other solid fat.

- ✦ Shortenings like Crisco contain emulsifiers that help batters and doughs come together faster.

- ✦ Unlike butter, shortening is 100 percent fat.

- ✦ Although they're made from vegetable oils that are high in unsaturated fats, shortenings are partially hydrogenated, which makes them mostly saturated.

Shortening is amazingly useful stuff because it's everything that butter isn't. It remains plastic between 98°F and 110°F, which means that shortening will be workable when butter would be either rock hard or soup. Shortening is 100 percent oil…well, not quite. About 10 percent of shortening's volume is air, or—to be more precise—nitrogen, which doesn't hasten oxidation the way air does. This is good news for bakers because that air can be used to leaven foods.

The question is: if vegetable shortening is made from vegetable oil, what makes it solid? Read the ingredient list on the side of a can of Crisco and this is what you'll see:

Ingredients: Partially Hydrogenated Soybean and Cottonseed Oils, Mono- and Diglycerides.

Now that doesn't sound very good, does it? Well, let's break it down.

Remember that the real difference between the three different kinds of triglycerides comes down to how many seats are empty on the tram and how many are stocked with hydrogen. Both cottonseed oil and soybean oil have quite a number of empty seats on their trams, which explains why they're liquid at storage and use temperatures. To make them solid we've got to find a way to get hydrogen on board. The answer: hydrogenation.

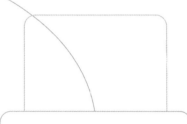

> **FACT**
>
> ✦ ✦ ✦
>
> Why do they call it shortening? Because fats (all fats, really) shorten gluten strands by lubricating them so they can't grip each other.

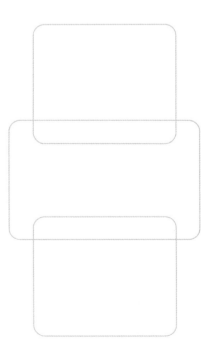

My shortening of choice.

First you heat up oil, then gurgle a lot of hydrogen through it in the presence of a catalyst (nickle), which speeds the reaction time and voilà, you've made yourself a solid fat—albeit one that has lost any of the possible health benefits of having been polyunsaturated in the first place. And what may (repeat "may") be worse is that the hydrogens that were forced onto the fatty acids don't connect at quite the same angle as natural fats. These new fatty acids are called "trans fatty" acids and they may be even worse for us than saturated fats. So if you've been using shortening as a way to avoid butter you may be wasting your time.

Okay, so now we're down to the mono- and diglycerides and they are exactly what you'd think. Instead of having three trans fatty acids, they only have one or two. These substances aren't fats per se but they are lipids, the larger chemical family that fats belong to. So why are they here? Because although the transfatty part of the molecule hates water as much as any self-respecting fat would, the rest of the molecule, the glycerol (or glycerin) likes water just fine. These substances are happy to stick one end in oil and the other in water. This means they can hold together two substances that don't ordinarily get along and that makes them emulsifiers, which are very valuable things indeed.

Although it's not listed in the ingredients, shortening is about 10 percent air, which means it can provide leavening without being creamed with sugar.

Basically, shortening can do everything that butter can do, and do it better—except for three things: it doesn't brown well; it tastes like, well, nothing; and in "laminated" doughs (like puff pastry), well, it's just a mess—that's because, with a melting point just above normal body temperature, it doesn't have a good mouthfeel.

Oils

Unlike butter, oil is all fat…that's it. No proteins, no solids, and unlike shortening, no tiny nitrogen bubbles. Just fat. Most of us have at least a few oils hanging around the kitchen, olive oil and maybe a flavored oil like chili oil. But when it comes to baking, we're mostly concerned with "vegetable oil." This term takes in a pretty wide range of products: soybean oil, peanut oil, cottonseed, corn, canola, and safflower oils.

WHY WORKING WITH SHORTENING IS EASIER

◆　◆　◆

Hydrogenated vegetable shortening has a melting point of 98° to 110°F. It remains pretty stable up to that temperature, that is, its texture doesn't change much. One minute it's shortening, and the next it's liquid fat. This is in contrast to butter, which begins to melt around 82°F and doesn't become a liquid until it reaches around 96.8°F. Since there are fewer factors to consider, shortening is more predictable and recipes that call for its use more consistent. That said, butter tastes a heck of a lot better (shortening doesn't really taste like anything until it goes rancid) nor does shortening assist browning, as butter can because of the milk solids and other proteins it contains.

Since they can't trap bubbles and don't contain any water, oils don't do any leavening whatsoever so products that use them usually have an uneven, coarse texture, which I, for one, like quite a bit. Since they're liquid at room temperature, they do however contribute a sense of moistness, despite the fact that they don't actually contain anything wet.

Oils are used in nearly all applications calling for the Muffin Method.

Oils, especially polyunsaturated oils, go bad quickly when exposed to air or light, so keep them in a cupboard in a cool place or buy oils in tins. I've got a big metal squirt soap dispenser that I keep oil in for squirting into sauté pans...but then I'm weird that way.

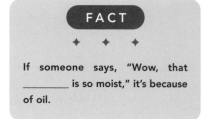

FACT

✦ ✦ ✦

If someone says, "Wow, that _____ is so moist," it's because of oil.

FATTY ACID PROFILES OF POPULAR OILS

OILS	Saturated %	Polyunsaturated %	Monounsaturated %
Canola	6	32	62
Almond	9	26	65
Safflower	9	76	15
Hazelnut	11	14	75
Sunflower	12	66	22
Corn	13	62	25
Grapeseed	13	70	17
Olive	14	10	76
Soybean	14	61	25
Walnut	16	56	28
Peanut	17	35	48
Sesame seed	18	41	41
Avocado	20	11	69
Cottonseed	26	53	21
Palm kernel	83	5	12
Coconut	89	3	8

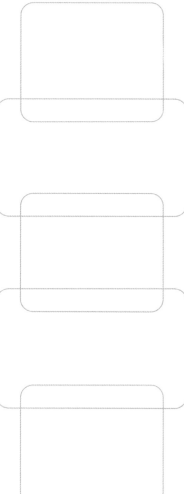

FACT

◆ ◆ ◆

The word "milk" comes from a very old Indo-European word *melk*, which refers to the action the hand makes when "milking" an animal.

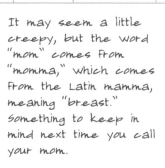

It may seem a little creepy, but the word "mom" comes from "momma," which comes from the Latin mamma, meaning "breast." Something to keep in mind next time you call your mom.

MILK (PROTEIN, FAT, WATER)

If you're reading this book, odds are you're a mammal. Our animal kingdom team is so named because the females of our kind have mammary glands that produce milk, which is used to feed our offspring. Since it's designed to provide the sole nourishment for a sophisticated carbon-based life form, milk is understandably complex stuff. Humans, especially humans of Northern European extraction, are unique among mammals because we tend to continue drinking and cooking with milk long after we are weaned— a fact that a lot of other mammals on this planet find revolting. Of course, it's not that we continue to drink human milk, now that would really be creepy. Nope, we drink milk from cows.

We drink milk from cows because, when properly managed, cows make a lot of the stuff. The stuff they make is more than 85 percent water, but the other 15 percent is unique and powerful. First there's milk sugar, or lactose; two different kinds of protein; fat; and a very interesting collection of salts that not only contribute flavor but promote protein coagulation in eggs.

The Things We Do to Milk

Louis Pasteur (one of the 10 most important people in the history of food, if you ask me) figured out that microbial forces were responsible for the spoiling of milk. The answer: heat. The process, pioneered by Pasteur, bears his name. Very few Americans have ever tasted milk straight from a cow because raw milk cannot be sold to consumers in this country. In fact, the only thing that raw milk can be used in is cheese that is aged at least six months before it's sold.

Personally I'm against this law because it ties the hands of American cheese makers, who could kick France's collective butt if they only had the chance. Alas, now that mad cow disease has broken out on U.S. soil, I fear the law will never change, despite the fact that the causative agent isn't carried in milk.

Pasteurization. There are several methods of pasteurization. The law requires (at the very least) that milk be heated to 145°F for 30 minutes. This is okay by me because at this temperature many of milk's true (and subtle) flavor compounds survive, while the majority of the nasty little bugs don't. Such milk is, however, almost impossible to find on grocery shelves because we Americans want our milk (as well as our beef, salmon, and food in general) dirt cheap. Sensing that we won't shell out a few more dimes for a gallon of milk that actually tastes like something, milk processors opt instead for something called HTST or "high-temperature-short-time" pasteurization, which takes only 15 seconds at 161°F. Of course, at these temperatures, the resulting milk tastes a lot like white water—but hey, that's industry.

If, under pressure, you're willing to jack that temperature to 280°F for even two seconds, you have UHT, or ultra-pasteurized milk, which tastes even less like anything. That said, if packed in an antiseptic carton, UHT milk can actually be stored (unopened) at room temperature for several months.

Homogenization. In its natural state, milk contains fat globules in a wide range of sizes. Given time, the larger ones find each other, join up, and—being less dense than the surrounding fluid—float to the top. This substance is called cream. To prevent this separation, whole milk can be pushed under pressure through tiny holes. This forces the fat globules to break up into a smaller, more uniform size.

Proteins and natural emulsifiers that don't mind bonding to water and fat immediately surround the new mini-globs. These substances essentially laminate the fat globules, preventing them from rebonding to form large globules that would then reseparate to make cream.

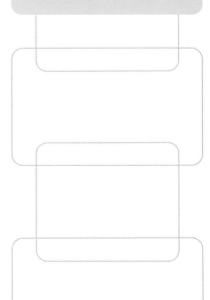

FACT

♦ ♦ ♦

Milk looks white for the same reason that a meringue looks white. It is composed of very small orbs—only instead of bubbles they are amalgamations of protein and calcium called "micelles." Light can't pass through them so white is reflected back. Light reflects off of milkfat globules the same way. That is why skim milk, which lacks these, often appears pale blue.

The process reminds me of making a salad dressing. If you just shake up oil and vinegar, it separates again within minutes. If, however, you put the two in a blender along with a little dry mustard, which contains small amounts of the emulsifier lecithin, and buzz it up, you'll have a very creamy looking liquid that will remain exactly the same for days, if not longer.

If the milk product is sweetened, this must be increased by 5°F.

Fat Levels in Milk

Since it would be impossible (or at the very least, terribly difficult) to remove the exact percentage of fat necessary to create say, 2-percent or 1-percent milk, most American milk is completely skimmed. All the fat is removed and then a specific percentage of fat is added back in, depending on what the market demands at any given time. Milkfat percentages and pasteurization times are as follows:

PRODUCT	Minimum fat	Pasteurization (not UHT)
Whole milk	not less than 3.25 percent	161°F for 15 seconds or 145°F for 30 minutes
Buttermilk	not less than 3.25 percent	same
Reduced-fat milk	at least 25 percent less total fat than whole milk	same
Lowfat milk	maximum of 3 grams or less total fat	same
Skim milk	less than 0.5 grams of total fat	same
Heavy cream	not less than 36 percent milkfat	166°F for 15 seconds or 150° F for 30 minutes
Light whipping cream	not less than 30 percent milkfat	same
Half-and-half	not less than 10.5 percent milkfat	same
Light cream	not less than 6 percent milkfat	same
Sour cream	not less than 18 percent milkfat	same
Reduced-fat sour cream	minimum of 25 percent reduction from full-fat	same
Lowfat sour cream	6 percent or less total fat	same
Yogurt	not less than 3.25 percent fat; not less than 8.25 percent milk solids	
Lowfat yogurt	not less than 0.5 percent or more than 2 percent fat; not less than 8.25 percent milk solids	

Other Dairy Products That Are Useful in Baked Goods

Evaporated milk and **sweetened condensed milk.** These canned products are just what they sound like. Evaporated milk is cooked until both the fat and nonfat milk solids are twice the original level. Sweetened condensed milk is even "drier" and has sugar added. Both products have a distinct caramel flavor that I really like. I often sneak sweetened condensed milk into recipes as a replacement for heavy cream and sugar (see the cheesecake recipe on page 314). I also put it in coffee from time to time.

Buttermilk. I have a really, really old recipe of my grandmother's that I tried to make for years. It was for a loaf bread, and it just never turned out. After poring over the recipe time after time, my mom finally pointed out the problem. The recipe called for buttermilk. Real buttermilk, that is—the liquid left over after the churning of butter. Today's buttermilk is little more than lowfat milk to which lactic acid bacteria have been added. These bugs produce a distinct acidic twang, and they thicken the milk noticeably. Since I didn't have raw milk, I couldn't make buttermilk and had to toss the recipe. Modern buttermilk remains one of my favorite ingredients, though. It contains enough acid to act as a leavener when paired with baking soda and its flavor is distinctly, well, tangy. Buttermilk—full-fat buttermilk—is always in my refrigerator.

Sour cream is light cream inoculated with lactic acid bacteria. In the United States, sour cream must contain a minimum of 5 percent lactic acid. It can often be interchanged with yogurt and sometimes even buttermilk, but don't try sliding in lowfat for full-fat—it just doesn't work.

PRODUCT	Maximum moisture	Percentage of milkfat by weight
Sour cream	55 percent	33 percent (minimum)
Reduced-fat cream cheese	70 percent	16.5 to 20 percent
Lite cream cheese	70 percent	0.5 to 16.5 percent
Neufchâtel cheese	65 percent	20 to 33 percent

Obviously a lot of play in the definition of "lite."

THE PROBLEM WITH MODERN MILK

♦ ♦ ♦

Whether it's wet or dry, bacteria love milk. Back in the days before high-heat pasteurization, the bacteria that came with the milk from the cow would slowly sour it. This sour milk or "clabber" was used much as buttermilk is today. But these bacteria are no longer present in milk, which means any bacteria that comes along can jump in with no competition. As a result, today's milk doesn't sour—it rots into a nasty, stinky mess. Refrigeration can slow this reaction considerably, as will storing the milk in the dark, or at least in an opaque container. Ultraviolet light can damage various structures in the milk, robbing it of nutritional value and creating off flavors. That's why most gallon jugs are frosted, or even better, yellow. Nostalgic though they may be, clear bottles are only suitable for a day or two of storage. Cultured products like sour cream and buttermilk have much longer shelf lives because their pH is far too low for most microbes to dwell comfortably.

Cream cheese was created in 1872 by a dairyman in upstate New York who was attempting to recreate a soft, unripened cheese from Normandy called Neufchâtel using cream rather than whole milk. When the cheese started to catch on with city folk downstate, he decided to name it "Philadelphia" cream cheese because at the time, Philly was considered the classiest town in America.

The trick to working with cream cheese is to beat it soft before adding any other ingredients. If you don't, whatever batter you're building will always be lumpy—always.

Dry milk powder is made by removing all, or almost all of the water in milk. It's convenient because it lasts for months or even years if properly sealed. It isn't something that most home cooks keep around, and if you've ever tasted a glass of the stuff (when mixed with water), that's understandable. It's nasty—or, to be more exact, unnatural. But, in baked goods this goes unnoticed. Dry milk powder can be used as is where its protein and sugars are required or can be mixed with water and used as a straight replacement for milk (just follow the mixing instructions on the box or pouch). Unless I'm making something like a custard, I prefer to use powdered rather than fluid milk in recipes. I also always add a quarter cup or so to ice cream mixes. I'm not sure how to explain it, but I think the lactose helps create a smoother texture.

If you've got the answer to that one, email me please.

Milk's Job in Baked Goods

Since milk contains everything from sugars to fats to proteins to emulsifiers to stuff we haven't figured out yet, its effects on baked goods are profound and unilateral. Its complex chemistry helps stave off staling, softens crusts, and—of course—provides flavor.

If the fat content is high enough, dairy also provides a very nice foam. Whipped cream is structurally a lot like whipped egg whites but instead of protein, the walls are reinforced with clumps of fat. Cream is much easier to whip when it's cold so chill your bowl and whisk/beater. Unlike egg whites, which increase six to eight times in volume when whipped, cream usually only doubles in volume and it doesn't hold forever. But if whipped cream collapses, it can be rewhipped—as long as it's cold.

LEAVENERS

A leavener is either a bubble or something that blows a bubble. Bubbles are the empty spaces that become the rooms in a baked good. Tear open a muffin. See the open spaces? Leavening did that. Leaveners come in three forms: physical, chemical, and biological.

Physical Leaveners: Air and Steam

Since they rarely appear in ingredient lists, it's easy to forget about air and steam, and yet they do most of the leavening on this planet. In fact, some devices, like those built from *choux paste* and *popovers* soar to considerable heights on the expansive natures of air and steam alone. Despite the fact that they contain some chemical leaveners, most cakes are actually lifted by the air integrated through the Creaming Method.

Even in cases when CO_2 does most of the heavy lifting, air plays a crucial role in leavening. When a batter is mixed or a dough kneaded, tiny bubbles are worked into the product. The bubbles are then blown up by the gases that have been produced by the chemical leaveners. Without these seed bubbles, all baked goods would have the open, coarse texture of a muffin...or worse. Why? Because the gas would just as soon blow up one big bubble as millions of little ones.

As for steam, it is the universal leavener because all baked goods contain water either in the form of milk, eggs, the water in butter, or just plain water. Steam is an amazingly efficient leavener because as water changes to vapor it expands in volume more than a thousand times. No matter what other leaveners are in use, steam and air will be at work too.

Chemical Leavening

The more acidic a substance is, the lower it falls on the pH scale, which is essentially a rating of its hydrogen ion content. Any solution with a pH under 7 is considered to be acidic. If a solution's pH is above 7, it's called an alkali or base. Although acidic ingredients are very common in the kitchen, bases are not because they usually have a soapy or

Citrus, buttermilk, yogurt, sour cream, vinegar, syrups like molasses and honey, brown sugar, and chocolate cocoa are all acidic to one degree or another.

Dispelling a Leavening Myth

MOST BAKERS ASSUME THAT leaveners create the bubbles that lift their baked goods. Actually, this is only true with goods that have relatively course textures, like muffins, or pâte à choux-based devices, or popovers, which have interiors like this.

In the case of cakes and yeast-leavened breads, the bubbles are actually "planted" by the baker through mixing and kneading. In the case of cakes, which have fine, tight textures, the bubbles are created by the cutting of sugar granules into solid fat, a crucial step in the Creaming Method, as we shall soon see.

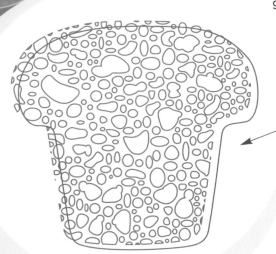

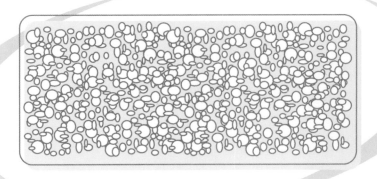

Fine textures are created in yeast breads by kneading and by the folding that follows the first and (occasionally) the second rise. This folding breaks up the large bubbles blown during the first rise so that they may rise again. If the baker is careful, his or her bubbles can literally multiply right alongside the yeast that exhale into them. As you'll see as you read on, most of the time the bubbles in question are born when you aerate the batter/dough during mixing/kneading. When it comes to yeast breads, cookies and cakes, these gases simply blow up bubbles that were invited to the party during mixing and/or kneading. It's not that CO_2 (or the alcohol vapors given off by yeast) don't have enough oomph to lift a dough or batter, it's just that left to their own devices they'd create just a few really big bubbles and the inside of your cake would look like this rather than this.

Since the walls of these bubbles set during baking—thus creating the texture of the finished good—you can see how these starter or "seed" bubbles could be kinda important.

chemical taste. When bases and acids come together in the presence of water, an interesting and quite fortuitous reaction occurs: they create a neutral salt and release a fair amount of carbon dioxide (CO_2), which can be used to leaven a wide range of baked goods.

Since there aren't too many bases hanging around the kitchen (ammonia and egg whites are the only ones I can think of), we require the services of a special additive to mix with our acids: **baking soda.**

Pearl Ash, or potassium carbonate extracted from wood ash, was first used as a baking soda in America around 1790. Although effective, it reacted with fats in the food, forming a kind of soap that didn't taste very good. Eventually it was replaced by sodium bicarbonate ($NaHCO_3$) aka bicarbonate of soda, aka bicarb, aka baking soda, which can still produce a soapy flavor, unless it's counteracted by acid. Most of the world's baking soda is made from vast trona deposits found in mines in Wyoming and dry lake beds in California. Who knew?

With acidic ingredients and baking soda on hand, you can produce CO_2. But there's a catch: in order to produce the maximum amount of gas, the acid and base must be balanced—and unless you've got a chemistry degree and big ole pad of litmus test paper, that's tough to do. Besides, you might want the acidic ingredients in your dough or batter to stay that way for the sake of flavor.

Let's say you make up a batch of buttermilk biscuits and you want them to rise nice and tall like the ones in the television commercials. So you decide to up the soda from ¼ to ½ teaspoon. Your biscuits come out nice and tall, yes, but upon consumption you note that the flavor is rather flat and boring. The problem is that the soda neutralized all the acid in the buttermilk, creating more CO_2 but leaving nothing behind for flavor. The solution? Well, you could add a pinch of **cream of tartar** to up the acidity of the batter.

Cream of tartar, or potassium hydrogen tartrate, is a pure and all-natural acid that's harvested from the inside of wine barrels in the form of tiny crystals. If you've ever opened a bottle of red wine and noticed tiny sparkles reflecting off the bottom of the cork, then you've seen tartaric acid crystals. If you had enough such corks, you could harvest

Alkalies are commonly extracted from the ashes of burned plants. If you want to see how corrosive they can be, let the bottom of your grill fill up with charcoal ashes, then leave it in the rain a few times. The ash solution will eventually eat a hole in the bottom of the grill. Or you can rent the movie Fight Club and watch the chemical burn scene. . . .

your very own cream of tartar…if that sounds like fun to you. Anyway, cream of tartar is useful because despite its considerable acidity, it doesn't have much in the way of flavor and it's a powder.

So, if cream of tartar is an acid and baking soda is a base, why not put them together and sell them as **"baking powder"**? Good idea.

Baking powders contain baking soda as well as an acid with which it can react. Since all it takes is a wee little bit of moisture to start the reaction, cornstarch is also added to absorb atmospheric moisture. Although there are several baking powders on the market (as opposed to baking soda—as far as I know there's only one of those) you could make your own thusly:

SINGLE ACTING BAKING POWDER

INGREDIENT	AMOUNT
Cream of tartar	2 parts
Baking soda	1 part
Cornstarch	1 part
Combine thoroughly and store in cool, air-dry place for up to 6 months.	

One must remember that this is a "single-acting" baking powder, meaning that it will give off a bunch of CO_2 within moments of being mixed into a liquid, then nothing. So although this would probably work fine in cookies or biscuits, it's not very useful in low-viscosity batters that can't hold on to bubbles such as cakes and quick breads. For these I prefer a commercial "double-acting" baking powder, which contains two or more different acids, one that dissolves and gives off CO_2 immediately upon mixing and another that doesn't give off gas until it's heated. That gives the baked good in question an opportunity to catch another lift before setting. As of this writing, all baking powders available at the consumer level are double-acting.

FACT

✦ ✦ ✦

All baking powders available in this country give off the same amount of CO_2: 12 percent by weight.

FACT

✦ ✦ ✦

Once opened, a can of baking powder only has a lifespan of six months.

pH of Common Stuff

CONCENTRATION OF HYDROGEN IONS	pH	Ingredient
10,000,000	pH 0	battery acid
1,000,000	pH 1	hydrochloric acid
100,000	pH 2	lemon juice, vinegar, gastric juices
10,000	pH 3	citrus juices
1,000	pH 4	ripe tomato
100	pH 5	black coffee
10	pH 6	saliva
0	pH 7	distilled water
1/10	pH 8	sea water
1/100	pH 9	baking soda
1/1,000	pH 10	milk of magnesia
1/10,000	pH 11	ammonia solution
1/100,000	pH 12	soapy water
1/1,000,000	pH 13	bleach, oven cleaner
1/10,000,000	pH 14	drain cleaner

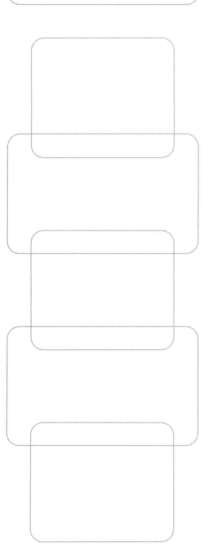

Depending on ripeness and other agro-factors.

Yeast: The Very Least You Should Know

✦ Yeast cells are alive...like sea monkeys. Treat them nicely until you're ready to kill them. That means giving them a warm, moist place to grow away from too much salt, sugar, or cleaning supplies.

✦ Instant yeast is superior to active dry. You can add it directly to the flour before mixing doughs.

✦ Slow rises in a cool place will produce better flavor and texture every time.

- Baker's yeast (*Saccharomyces cerevisiae*) is a specific strain of brewer's yeast, but don't get any ideas. Certain Native American tribes used to make a potent drink from the liquid run-off from fermenting doughs called "hooch" and the headache the next day was a thing of legend.

- To use instant yeast in a recipe calling for active dry, just substitute them one for one. It's not exact, but it's close enough.

- Without salt, yeast run wild, often overproducing to the point where the entire population burns itself out before the bread gets to the oven. Salt keeps the population under control.

Three of my favorite things on Earth—bread, wine, and beer—are all made possible by the metabolic action of a microscopic single-cell fungus called yeast. Up until a century or two ago, brewers produced all the yeast. They'd scrape foam from the fermenting beer tanks, dry it, and sell it to bakers. In fact, of the hundreds of known strains of commercial yeast, the one used in bread making most often is *Sacchcromyces cervevisiae* (Latin for "beer sugar").

The metabolic action we're speaking of is of course, fermentation. Yeast encounter sugar, yeast eat sugar (or break it down, to be more exact), and in turn produce carbon dioxide, alcohol, heat energy, and a host of flavor compounds. Bakers cherish yeast granules for their gassy nature and brewers for their ability to crank out ethanol.

Although bakers and brewers have been depending on yeast for a dozen millennia or more, until Louis Pasteur "discovered" yeast (along with a bunch of other things) back in the nineteenth century, they figured that foamy stuff in the bowl was a miracle. Brewers, in fact, called that yeast-laden foam "godisgoode."

It's important to note that since yeast cells lack the enzyme amylase, they cannot break the starch in flour down into sugar, which is a shame because it's sugar that they eat. To make up for this, most millers add amylase to their flours. If you cultivate your own wild sourdough starter, wild bacteria (called *lactobacilli*) that can break down

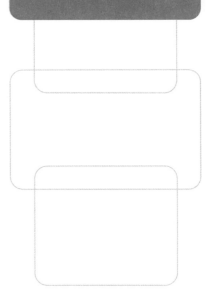

AMMONIUM BICARBONATE

◆ ◆ ◆

If you have a really old American cookbook or baking formulas from Europe you may run into this ingredient. Ammonium Bicarbonate is a unique leavener in that it can give off leavening gas without reacting to an acid or an alkali. All it needs is moisture and heat. The problem is, it reeks to high heaven. I've used it a few times (brought it back from Europe...Customs loved that) and the kitchen smelled so much of ammonia that I thought I'd have to evacuate the place. And, unless the good in question is relatively small and possessing a high surface to mass ratio (say a cookie or a cracker) this gas will accumulate in the product making it taste as bad as it smells.

FACT

✦ ✦ ✦

Strange but true: A healthy, well-fed yeast colony can produce an entire generation in five hours.

Compressed yeast blocks are only about 30 percent yeast. The rest is moisture.

sugars will no doubt drift into the party, which is why starters are such efficient leaveners. (For more on starters, see pages 232–235.)

Types of Yeast

These days yeast comes in several reliable market forms.

Fresh or cake yeast. Basically zillions and zillions of yeast cells compressed into blocks but still wide-awake and ready to go. Now some bakers swear by the fresh stuff, but I think it's more trouble than it's worth. It has to be kept wrapped and refrigerated and even then only remains viable for a couple of weeks. It's like adopting an old goldfish. No matter what you do, it's going belly-up in a week. So choose according to how soon you'll be using it.

Active dry yeast (aka the stuff in the little triple pouches). An oxymoron if ever there was one, these are essentially yeast mudskippers, which must be roused to action by "proofing." If you've ever executed a yeast-leavened recipe that called for sprinkling the packet of yeast over a bowl of warm water or warm sugar water, you have "proofed" yeast. If after a few minutes the mixture didn't foam up, you knew you were in possession of dead yeast. Of course, even if your mix foamed like a warm beer you'd still have dead yeast on your hands, because on average 25 percent of active dry yeast are permanently nonactive, if you get my drift. In and of itself, this wouldn't be a big deal, but dead yeast cells emit chemicals that can monkey with the outcome of the baked good. If you simply must use this kind of yeast, proof it in four times its weight of warm (105° to 115°F) water.

Another reason why I don't care for active dry yeast is that I don't like having to structure the mixing procedure around having to wake up a bunch of fungus. And if you get the water too hot, say 140°F, your conscience will have to bear the weight of zillions of fungal deaths.

Instant yeast (available in pouches, jars, and one-pound bags) is a hardy strain of dry yeast that's been broken into very, very small granules and then packed with vitamin C in the form of ascorbic acid. This really turns the yeast on upon waking. Although you could argue that this is nothing more than active dry version 3.0, the drying process that's used is more gentle than the traditional methods, so you don't have that minimum 25 percent mortality rate to fret about. Instant yeast is also meant to be mixed directly into the dry ingredients—no soaking or proofing first, which means no chance of killing it with too hot water. It also doesn't even mind being mixed with salt—while dry, that is. When using instant yeast, you should also be sure the initial dough temperatures are between 70° and 95°F.

Although both Fleischmann's and Red Star make instant yeast, I'm a SAF man and buy a pound at a time. Whenever I buy a bag I split it up so I've got a small working supply in the refrigerator and reinforcements standing by in long-term, air-tight storage in the freezer. By the way, the expiration date on the bag will always be a year from the date of manufacture, but kept very cold and dry you'll get two years out of it easy.

Instant yeast is sometimes marketed as "rapid rise" yeast, but that's not a true indication of what it does. Instant yeast doesn't actually rise faster than active dry yeast, they're just more robust, so they're more efficient…and of course 25 percent of them aren't dead.

Sourdough Starters

You can make good breads from commercial yeast, but if you're going to make great bread—really great bread—you're going to have to cultivate your own sourdough starter. Raising true sourdough starters is a bit like capturing and training any wild animal…fungus…you know what I mean. The down side is that you don't know where they are or what they might bring with them when they show up. Volumes have been written about sourdoughs. Those by Ed Wood are especially good, as are the starter cultures from around the world available from his www.sourdo.com.

The starter that I use—the one in the Straight Dough Method on page 232—is a cheat. It's not a true sourdough, but a live liquid culture created from commercial yeast. It's not as robust as a wild starter and it has a tough time lifting a dough all on its own. That means I have to bolster it with instant yeast whenever I use it. So why bother?

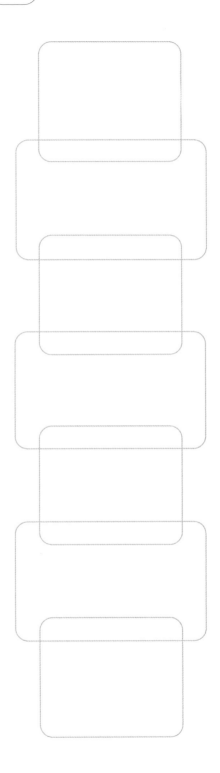

Flavor. It's very tangy and complex, and by using it I get more flavor and texture into my bread in very little time and with only one, or sometimes two rises.

Things That Affect Yeast Activity

Dough Temperature. Yeast speeds up and slows down depending on the temperature.

DOUGH TEMPERATURE	YEAST ACTIVITY
34°F	inactive
60° to 70°F	slow
70° to 90°F	best growth
More than 100°F	slows again
140°F	dead or dying

Salt. Old-fashioned bread procedures, designed for use with traditional active dry yeast, usually call for the salt to be added separately from the dry goods or even after the dough has been mixed. For me, this translates into lots of ruined bread because I often forget to add it. Why bother? Because yeast are a form of microbial life, and such life forms usually don't like to have a whole lot of salt around for the same reason that a slug doesn't like having it sprinkled on its back. Osmosis, the same force that makes brining possible, will see to it that if the concentration of salt gets high enough around the yeast, moisture will head out of the yeast and into the solution, eventually leaving the little bugger high, dry, and dead. However, this can be a good thing. Salt retards or slows yeast fermentation, which means that during a slow rise (which I prefer) you don't have to punch down the dough nearly as often. And since it holds on to moisture, salt will help your bread stay fresher longer.

Spices. Since most aromatic spices (cloves, cinnamon, nutmeg, and so on) act as antimicrobial agents, they can definitely slow yeast down.

Sugar. Although yeast cells usually go to town on sugar, high amounts of it will actually slow them down, which is why many sweet yeast dough formulas call for adding part of the sugar after the first rise.

Chlorine. It kills stuff…real well. That's what it does—that's *all* it does, so when I bake with yeast, I use filtered or bottled water. If you can't manage either, draw the water and let it sit open for six hours so that the chlorine can dissipate.

THE SECONDARY INGREDIENTS

A Few Words on Secondary Ingredients

While certainly important, there are many ingredients that I don't consider engine parts—that is, they enhance the baked good, but they don't define it. Although there are hundreds of such ingredients, my focus here is on those commonly called for in the applications to come.

Salt

I don't know too many baked goods that don't include at least some salt. Besides contributing its own flavor, salt can "complete" flavors as well as "turn them up." In other words, salt makes things taste more like themselves. In even small amounts, salt can make chocolate taste more like chocolate or a tomato more like a tomato. Salt is the only single seasoning our tongues are programmed to taste. It's that important. Don't believe me? Bake bread without salt in it…heck, even the dog won't eat it.

But salt's influence on baked goods goes well beyond flavor. Salt strengthens gluten, and provides structure in bread making. Salt is partially responsible for the darkening of crusts during baking. Since it's hygroscopic, salt helps prevent staleness. As an agent of osmosis, salt limits yeast activity. (If it can dry a ham, just imagine what it can do to a unicellular mushroom.)

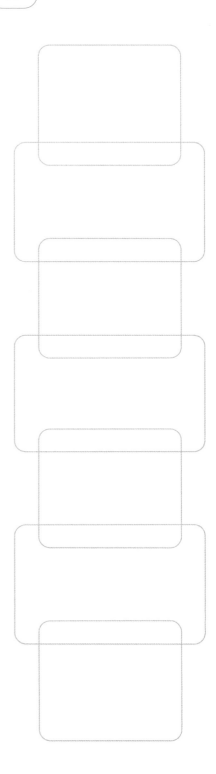

A FEW MORE WORDS ON VANILLA

◆ ◆ ◆

I really resent the use of "vanilla" to describe something boring. Vanilla is not boring, either botanically or culinarily speaking. Consider the following:

Vanilla is a plant, an orchid in fact, and the "beans" are its pods. Although only two actual varieties—Bourbon and Tahitian—are used commercially, the growing location and processing method greatly determine the final flavor. In that respect, vanilla is exactly like coffee.

Vanilla is not an easy thing to process. The plants are tricky to raise and pollinate, and once they produce pods they take the better part of a year to ripen. They must be picked by hand and dried or "cured" in order to develop flavor and aroma (just like coffee and chocolate).

Although Bourbon pods from Madagascar are the most popular in the United States, and sweet Tahitian pods are big in Europe, I prefer Mexican-raised Bourbons myself, whether I'm using them whole, pulp-only, or extracted.

Extracts

Extracts are the most common processed flavorings found in baked goods. Extracts are made by chopping and soaking ingredients such as herbs, nuts, berries, and other ingredients containing alcohol-soluble compounds (usually oils) in ethyl alcohol or a combination of alcohol and water. And the most commonly used of these is vanilla extract.

In the United States, "real" vanilla extract must contain 35 percent alcohol and 13.35 ounces of vanilla beans per gallon. Double strength extracts contain the same amount of alcohol, but twice the beans. That doesn't mean that double strength is better, but it does mean that it tastes different than single strength. Imitation vanilla extract is made from wood and other stuff. Vanilla flavoring, on the other hand, is a combination of imitation and pure extracts.

Artificial extracts count on combinations of chemicals to replace the flavor and aroma of the "real" thing. I use artificial vanilla extract in baked goods because I think it tastes just as real as the real thing, if not more like the real thing than the real thing, which is confusing because, technically speaking, imitation vanilla extract, which contains vanillin and coumarin, are chemically identical to the "real thing." In other words, it is possible to manufacture a vanilla extract that is as real as actual vanilla.

I use the fake stuff whenever vanilla is a supporting player, rather than the frontline flavor—I'd use the fake in brownies, but not in vanilla ice cream.

Real Vanilla Beans

If a whole bean is called for in a recipe it's usually split and the pulp is scraped out. Sometimes the bean is then chopped and included in the mixture but since it's tough as leather it's always strained out. I prefer to save the shell of the pod for vanilla sugar (see page 113).

Liqueurs

Liqueurs are basically weak extracts with a lot of sugar added. Fruit liqueurs, nut liqueurs, chocolate and coffee liqueurs, and anise liqueurs are the most commonly used

> Wouldn't Dr. Tyrell love that?

in baked goods, but occasionally you'll see an application call for an herbal or "bitter" liqueur such as Benedictine or Campari. When I want to use a liqueur rather than an extract, I double the amount called for. I also subtract a teaspoon of sugar from the recipe for every tablespoon of liqueur added.

It should be noted that not all liqueurs containing the same flavor share the same taste. For instance, Cointreau and Grand Marnier are both orange liqueurs, but they taste very different from one another. Ditto Tia Maria and Kahlúa.

A Few Words on Chocolate

Chocolate in all its guises hails from the seeds that form inside the pod that grows upon the spindly cacao tree in South America. These seeds, or "beans," are harvested and then allowed to dry and partially ferment in open air. After a slow roasting, the outer hulls are removed to reveal the inner nibs. The nibs are then rolled under heavy stone or metal wheels to produce a brown paste that's called *chocolate liquor*, despite the fact that there's no alcohol involved. From the time of the Maya to the eighteenth century, this substance was simply mixed with a few spices, frothed into water, and served. This bitter but stimulating brew became so popular with the Spanish settlers in Mexico that women started taking gourdfuls of the stuff with them to Mass just to stay awake through all that steamy Latin. When a powerful bishop in Chiapas spoke out against the practice, he was assassinated with poisoned cocoa.

Emperor Montezuma was said to drink 50 gourds of the stuff a day.

Chocolate remained primarily a beverage until 1828, when a Dutch chocolatier named Conrad Van Houten devised a hydraulic press that could separate the cocoa solids, or cake, from the fat, or cocoa butter. Pulverize this cake and you've got natural cocoa powder, which is still common in the United States. Natural cocoa is easy to spot because it's brick-red. With a pH of 5.2, it is fairly acidic. Hoping to make his cocoa powder more palatable, Van Houten added alkalies to it, and that mellowed out the flavor. It also darkened the color into something more chocolaty. This type of cocoa, which is very common in Europe, is called *Dutch-process* or *Dutched* cocoa.

Some cooks find that Dutched cocoa mixes more easily with liquids than natural, but I've never seen anything to suggest that. What does affect the solubility of cocoa powder is the percentage of cocoa butter that's still clinging to the granules. Since that differs from brand to brand, you'll just have to shop around until you find one that makes you happy. My preference is Droste, but in a pinch, I still use good ole Hershey's.

Baking chocolate, or unsweetened chocolate, is nothing more than chocolate liquor (that is, cocoa powder with cocoa butter) stabilized in cube or bar form. It's often used alongside cocoa powder in brownies and fudge because the cocoa butter is extremely hard at room temperature but completely liquid by the time it hits 95°F. In firm baked goods like brownies, this can be a good thing. However, because of its high melting point, cocoa butter is not considered a "tenderizer" in most baked goods.

Sweetened chocolate, semisweet, bittersweet, milk, and dark chocolate all contain different percentages of cocoa solids, sugar, and other ingredients such as vanilla, milk solids, and emulsifiers. Although I prefer dark European-style chocolates for baking, these terms are, for the sake of most baking, synonymous. Now you chocolate fans, don't get all fired up. I'm not saying they *are* the same I'm just saying that they *act* the same in recipes. Part of how I decide which chocolate to use depends on the baked good. If I'm making chocolate chip cookies, I use good ole bagged morsels; if the chocolate is to be melted in a double boiler before being added into a batter, I use the best chocolate I can get my hands on. My general rule is, when it comes to sweetened chocolates, use what you like to eat. Besides, chocolate's another book.

Remember, top-notch ingredients can give you an edge but nothing beats good technique.

Baking Under Pressure

I had the opportunity to meet an astronaut once and I asked her what would happen if she took her helmet off in outer space. She told me without hesitation that nothing could possibly prompt her to do this. Sensing immediately that astronauts are very technical people, I rephrased the question to include acts of accidental removal. She assured me that, despite Hollywood's portrayals, one's eyes would not bug out nor would one's head explode, though your eardrums would. That's because any air space expands very quickly. She went on to tell me that she had herself experienced the phenomenon (thankfully momentarily) in a vacuum chamber and that it was like eating a Zot. Those of us who consumed mass quantities of candies in the 1970s remember well the intense fizzing Zots created on the tongue. The reason this happens to ill-fated astronauts is that, in the absence of atmospheric pressure, spit instantly boils. Not because it's hot, but because there is no longer any pressure to hold the water molecules together.

My point in telling you all this is that, while most of us think that the boiling of water has a lot to do with heat, it's really about pressure. At sea level, the surrounding atmosphere exerts 14.7 pounds per square inch of pressure on everything from every angle. Think of 14.7 pounds as the weight of the sky directly above you right now, or on a pot of water, or on the water in a cake batter. If we want to convert that water to a vapor form we have to do one of two things:

◆ Expose it to so much heat energy that the pressure inside the solution will eventually surpass the surrounding atmospheric pressure and the water will "boil," or

◆ We can climb Mt. Everest, or better yet, take our water up in a balloon capable of rising into the stratosphere, where very little heat energy is needed to boil water.

If the condition were to continue, your blood would boil, too. . .just before you died from lack of air.

This is a gross simplification, but t'will do for now.

Of course, if we go the other way, we experience the inverse. By applying approximately 15 psi, the temperature inside a pressure cooker can elevate to 257°F, which is very impressive indeed.

Another issue to consider regarding pressure is what happens to air inside a bubble. A balloon half blown up at sea level will expand to full in a few thousand feet and explode soon thereafter. That's because nature hates a vacuum. Nature would like to expand the air inside the balloon so that the pressure inside it equals that outside it. That's why Goldfinger's fat butt got sucked right out that little-bitty airplane window at the end of the third 007 movie.

What's all this got to do with baking? Well, baking is to a large extent about managing water and bubbles, so the answer is "a lot." And since most published recipes are tested at or near sea level, a baker in Denver, Colorado, or in base camp on Mt. Everest might experience some frustrating complications. This is especially true of light, air-filled batters like those in angel food cake and meringue and less true of dense batters like brownies and custards. What can be done? Ideally, new recipes should be devised at those elevations, but you can also tinker with the pressure sensitive factors inside your baked goods.

Grease pans well. Although science has yet to explain this sufficiently, baked goods stick more at high altitude. Go figure.

Adjust your leavening. Gases expand more and faster as the pressure is reduced, so the amounts of chemical leavenings and yeast used should be reduced at higher altitudes. If most of the leavening is to be provided by air that's physically incorporated, as in creaming, the baker should attempt to work less air into the product.

Cut back on both sugar and fats. This will slightly reduce the rate at which the cake will set, making better use of the leavening and the water inside the batter.

Raise the temperature. In order to capture the bubbles created by leavening and moisture, baking temperatures should be increased by 25°F above 3,500 feet to speed the coagulation of proteins and the gelatinization of starches. Also, adding an egg yolk will provide moisture and protein helping the structure set earlier and stay moister longer.

Reduce rising time. Rising times for yeast breads should be reduced slightly so that bread doesn't overproof. And when working with yeast, go with a high protein flour. A higher protein content will result in stronger gluten capable of stretching with quickly expanding bubbles.

HIGH ALTITUDE CONVERSION FOR BOILING WATER

ALTITUDE	°F	°C
Sea Level	212.0	100.0
2,000 ft	208.4	98.4
5,000 ft	203.0	95.0
7,500 ft	198.4	92.4
10,000 ft	194.0	90.0

HIGH ALTITUDE BAKING ADJUSTMENTS

FEET ABOVE SEA LEVEL	Baking Powder (REDUCE EACH t BY)	Sugar (REDUCE EACH CUP BY)	Liquid (FOR EACH CUP ADD)
3,000	⅛ t	½ to 1 T	1 to 2 T
5,000	⅛ to ¼ t	½ to 2 T	2 to 4 T
7,000 and above	¼ t	1 to 3 T	3 to 4 T

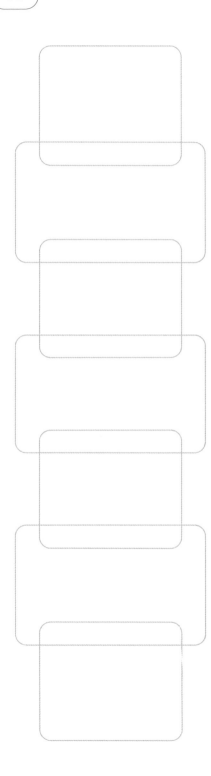

The Muffin Method

What's it do? It creates the uneven yet relatively tender interior characteristic of muffins, pancakes, and other quickbreads. With this method in your head, about 65 percent of the baked goods on earth are in the bag. To my mind this is a more precise method than creaming (which we'll discuss a little further on), because there are no peskily nebulous verbs like "cream."

Now, suppose you crack open a muffin and find something like this.

Would you think the strange meandering cavities you see here were formed by:

a) An adult muffin worm?

b) The expansion of gases?

c) A hunk of solid fat that melted during baking?

If you chose **a**, well, that's just sad. As for **c**, we assume that—being a muffin—this was assembled with liquid fat rather than solid, so no cigar. Therefore, the answer should be **b**.

So our next question is, why did gas do this?

a) too much chemical leavening?

b) over-mixing?

c) oven too hot?

Well, excessive leavening usually leads to a sunken or collapsed baked good because too much batter expansion takes place too soon. Either that or the batter is simply stretched too thin to support its own weight.

But a too-hot oven usually results in either a pointy top or lack of expansion. The outer shell sets too soon, and the gas never gets to do its thing.

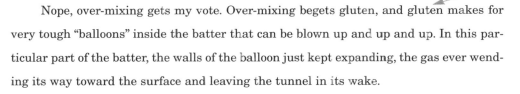

See page 40.

Nope, over-mixing gets my vote. Over-mixing begets gluten, and gluten makes for very tough "balloons" inside the batter that can be blown up and up and up. In this particular part of the batter, the walls of the balloon just kept expanding, the gas ever wending its way toward the surface and leaving the tunnel in its wake.

So when have you mixed just enough? Well, keeping notes and looking at the results will be your best guide, but here are two things that I do. Although many recipes call for stirring with a spoon or spatula, I use a whisk because it seems to get the deed done in the shortest time possible. However, if you don't sift your dry goods first, using a whisk may actually take longer. Most importantly, recipe instructions employing the Muffin Method will often say to mix the wet and dry ingredients together quickly and just until the batter "comes together." For me, this means dropping the whisk and walking away a good ten seconds before I think I should. There will still be lumps aplenty in the batter, but they'll bake out. If the batter is smooth, you have indeed gone too far.

What's the Deal with Sifting?

SIFTING IS MEANT TO ACCOMPLISH THREE THINGS: remove lumps, aerate dry goods, and effectively combine ingredients.

During storage, be it in bag or in canister, granules like flour settle closer and closer to each other, often forming lumps.

This is the main reason it's impossible to measure flour accurately without weighing.

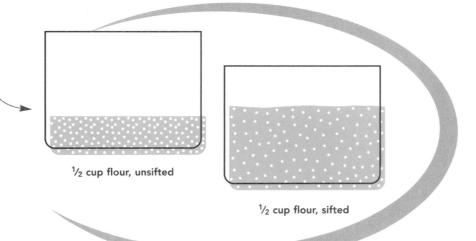

½ cup flour, unsifted

½ cup flour, sifted

Sifting decompresses the flour by working air into it and, as you might expect, granules surrounded by air are much easier to work into a batter—and that means less stirring, and that means less gluten. And that's a good thing, especially if you're making muffins or pancakes.

So sifting speeds the integration of flour into a batter by aeration and, although I don't have the lab tools to prove it, I firmly believe that sifting assists leavening too. As liquid moves into aerated flour, some of that air is going to be trapped as bubbles. I know this is true because if I make two sets of biscuits, following an identical recipe with the exception that I sift the dry goods in one and not the other, the sifted biscuits are always lighter—always. Granted, the fact that they mix quicker could mean less gluten production—and that

would mean the chemical leavening would have more room to work—but I'm not buying it. I think there are just more bubbles there to begin with.

Sifting is also a way to effectively combine dry ingredients. No matter how much you stir or whisk, it's just not easy to evenly distribute small amounts of baking powder, soda, or salt throughout large amounts of flour. Sifting always has been seen as a good way to ensure that this happens.

I think this may be true if you sift two or even three times with a traditional sifter, but, man, does that make a mess. Besides, I hate sifters—really hate them. The mechanisms always rust (even if you don't get them wet), they throw flour around, and they're real pains to store.

Nope, if a recipe calls for combining the dry ingredients, I dump them in the food processor and take 'em for a five-second spin. Then I just pour from the work bowl into the mixer bowl (or wherever they're going).

My rule is: always combine the dry goods in a food processor before adding them to the wet works—whether the formula/recipe asks you to sift or not. It's just a good policy.

My idea of a sifter.

Old-School Muffins

If you walk into a Starbu…um, a coffee place these days and order a muffin, what you'll get will be something between a carrot cake and a classic cupcake—that's not a real muffin. Real muffins should be coarser and less sweet than a cake, and although they contain enough fat to be tender, I wouldn't exactly call them moist. I would, however, call them delicious. This is an all-purpose formula that can take anything from berries to chocolate chips.

Hardware:

Digital scale

Dry measuring cups

Wet measuring cups

Measuring spoons

Food processor

2 large mixing bowls

Whisk

Rubber or silicone spatula

12-hole muffin tin

Disher

Instant-read thermometer or toothpick

Cooling rack

Notes:

Nuts, berries, chocolate chips, or any combination thereof—but no more than 1 cup of chips.

THE DRY GOODS

INGREDIENT	Weight		Volume	Count	Prep
All-purpose flour	303 g	11 oz	2$\frac{1}{4}$ cups		
Baking powder	7 g	$\frac{1}{4}$ oz	2 teaspoons		
Baking soda	6 g	<$\frac{1}{4}$ oz	1 teaspoon		
Salt			pinch		

THE WET WORKS

INGREDIENT	Weight		Volume	Count	Prep
Sugar	105 g	3$\frac{3}{4}$ oz	$\frac{1}{2}$ cup		
Vegetable oil			$\frac{1}{2}$ cup		
Whole egg	50 g	1$\frac{3}{4}$ oz		1 large	
Egg yolk	20 g	$\frac{2}{3}$ oz		1 large	
Plain yogurt			1 cup		

THE EXTRAS

INGREDIENT	Weight	Volume	Count	Prep
Bits and pieces		1 to 2 cups		
Baker's Joy or AB's Kustom Kitchen Lube for the tin				

✦ ✦ ✦ ✦ ✦

✦ Place an oven rack in position C and preheat the oven to 375°F.

✦ Prepare a muffin tin (see pages 180–183) and set aside.

✦ Assemble the Dry Goods, Wet Works, and Extras via the **MUFFIN METHOD.**

✦ Using the disher, drop the batter into the prepared tin. The cups should be full.

✦ Bake for 18 to 20 minutes, or until the muffin interiors hit 210°F or a toothpick inserted into the bottom of a muffin comes out clean.

✦ Remove from the oven and immediately turn the muffins on their sides so that steam can escape the pan. Skip this step and you'll have mushy muffin bottoms, which is okay if you like that kind of thing.

✦ Serve immediately or store in an airtight container for up to a week or until they taste gross, smell bad, or grow fur.

Yield: 12 standard muffins

To take the temperature of a baked good, use an instant-read thermometer. Grab your pot holders, then pop the target item from its pan momentarily and insert the thermometer in the bottom.

Mini muffins? I think not.

THE TRUTH ABOUT MUFFIN BATTER

✦ ✦ ✦

MUFFIN MYTH: Since the baking powder/soda starts releasing leavening immediately upon mixing, such batters need to be cooked immediately.

MUFFIN TRUTH: Muffin batter can handle a night in the chill chest as long as you don't stir it up before you pan it. The batter is viscous enough to hold the bubbles in as long as you don't go letting them out. In any case, I suggest portioning muffins into the pan with a disher like the one shown below. Cold batter is easier to portion than warm.

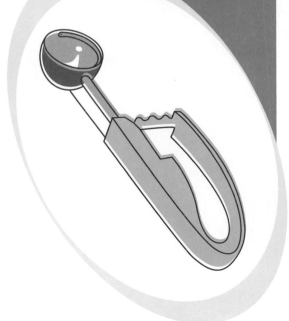

Hardware:

Digital scale

Dry measuring cups

Wet measuring cups

Measuring spoons

Food processor

2 large mixing bowls

Whisk

Rubber or silicone spatula

12-hole muffin tin

Disher

Instant-read thermometer
or toothpick

Notes:

(See page 186.)

Chocolate Muffins #7

Why number seven? Because in order to bring you the ultimate muffin recipe, we tested thousands of different ingredient combinations and mixing methods...well, okay, we tested seven. Some weren't bad; one even yielded a great chocolate cupcake, but this is a true chocolate muffin.

THE DRY GOODS

INGREDIENT	Weight		Volume	Count	Prep
All-purpose flour	270 g	9½ oz	2 cups		
Cocoa powder	92 g	3¼ oz	¾ cup		
Baking powder	9 g	¼ oz	1 tablespoon		
Baking soda	3 g	⅛ oz	½ teaspoon		
Salt	3 g	⅛ oz	½ teaspoon		

THE WET WORKS

INGREDIENT	Weight		Volume	Count	Prep
Sugar	263 g	9½ oz	1¼ cup		
Unsalted butter	113 g	4 oz	8 tablespoons	1 stick	melted and slightly cooled
Eggs	100 g	3½ oz		2 large	
Buttermilk	227 g	8 oz	1 cup		
Vanilla extract	9 g	⅓ oz	2 teaspoons		

THE EXTRAS

INGREDIENT	Weight		Volume	Count	Prep
Chocolate chips	177 g	6⅓ oz	1 cup		
Baker's Joy or AB's Kustom Kitchen Lube for the tin					

Notes:

✦ ✦ ✦ ✦ ✦

✦ Place an oven rack in position C and preheat the oven to 375°F.

✦ Prepare a muffin tin (see pages 180–183) and set aside.

✦ Assemble the Dry Goods, Wet Works, and the chocolate chips via the **MUFFIN METHOD.**

✦ Using the disher, drop the batter into the prepared tin. The cups should be full.

✦ Place in the oven and increase the oven temperature to 400°F.

✦ Bake for 18 to 20 minutes, until the muffin interiors hit 210°F or a toothpick inserted into the bottom of a muffin comes out clean.

✦ Remove from the oven and immediately turn the muffins up on their sides so that steam can escape the pan.

✦ Serve immediately or store in an airtight container.

Yield: 12 standard muffins

Hardware:

Digital scale

Dry measuring cups

Wet measuring cups

Measuring spoons

Food processor

2 large mixing bowls

Whisk

Rubber or silicone spatula

1-ounce disher

2 baking sheets

Parchment

Cooling rack

Chocolate Chip Cookie #10

Cookies are traditionally made via the Creaming Method, and you'll find other results of our experiments in that section. But this is my favorite chocolate chip cookie version.

THE DRY GOODS

INGREDIENT	Weight		Volume	Count	Prep
All-purpose flour	300 g	10¾ oz	2¼ cups		
Baking soda	6 g	<¼ oz	1 teaspoon		
Salt	6 g	<¼ oz	1 teaspoon		

THE WET WORKS

INGREDIENT	Weight		Volume	Count	Prep
Sugar	150 g	5⅓ oz	¾ cup		
Brown sugar	142 g	5 oz	¾ cup		
Unsalted butter	227 g	8 oz	1 cup	2 sticks	melted and slightly cooled
Egg yolks	40 g	1⅓ oz		2 large	beaten
Vanilla extract	57 g	2 oz	1 teaspoon		

THE EXTRAS

INGREDIENT	Weight		Volume	Count	Prep
Chocolate chips	340 g	12 oz	2 cups		

✦ Place oven racks in positions B and C and preheat the oven to 375°F.

✦ Assemble the Dry Goods, Wet Works, and Extras via the **MUFFIN METHOD.**

✦ Using the disher, scoop 24 portions onto a pair of ungreased or parchment-lined baking sheets. At this stage, the cookies can be frozen on the pans then moved to zip-top bags; they'll keep for up to 3 months.

✦ Bake for 15 to 17 minutes, or until golden brown. Let the cookies cool on the baking sheets for 2 minutes, then remove to a rack to cool completely.

Yield: About 3 dozen 1-ounce cookies

COOKIE SHEETS

◆   ◆   ◆

There are many different types of cookie sheets available, but I like the jelly roll pan best. Also known as a half-sheet pan in the restaurant trade (and a restaurant supply store is where you should buy yours), this is a 17½ x 12½-inch aluminum pan with a slightly flared lip. I use these pans for so much more than cookies and think that every kitchen should have at least three—one for each rack in your oven.

Oven Boost

WHY CHANGE THE TEMPERATURE OF YOUR OVEN once the goods to be baked are inside? Oven spring, my friends. That's what we call the big, steam-stimulated rise that a high-moisture batter gets when it first goes into a hot oven. The thing is, oven spring depends on direct heat moving up from the bottom of the oven to the bottom of the pan. Since ovens cycle on and off depending on what their thermostats tell them, the only way to ensure a hardy spring of heat is to manually turn the oven on again by changing the temperature when you insert your baked goods.

Some bakers prefer to preheat to a higher temperature and then back the temperature down when the item goes in; some would rather give the oven a boost. Whether I reduce or boost depends a good deal on the size and shape of the device in question. I usually boost small items, while I often reduce larger loaves. That might seem a bit curious, but a rising oven goes up to its final destination—good for small items. A falling oven starts high and gradually goes down—good for large items like loaves that require less heat as they set and dry.

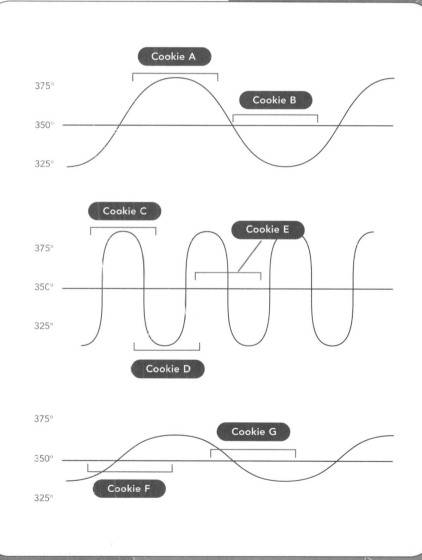

Hardware:

Digital scale

Dry measuring cups

Measuring spoons

Chef's knife

Cutting board

Food processor

3 large mixing bowls

Potato masher

Rubber or silicone spatula

9 x 5 x 2½-inch loaf pan

Instant-read thermometer
or toothpick

Cooling rack

Notes:

Banana Bread

Ah, poor banana bread—such a bad reputation. And it's a shame because few things toast better for breakfast. The problem is usually too much sugar, underripe bananas, and too much mixing! The bananas that go into this can be way gone—as in black—on the outside, that is. As long as there isn't fur growing on anything and nothing is oozing (the only food that's allowed to ooze is cheese), you're good to go.

THE WET WORKS 1

INGREDIENT	Weight		Volume	Count	Prep
Overripe bananas	340 g	12 oz		3 to 4	
Sugar	210 g	7½ oz	1 cup		

THE DRY GOODS

INGREDIENT	Weight		Volume	Count	Prep
All-purpose flour	220 g	7¾ oz	1⅔ cups		
Oat flour	35 g	1¼ oz	⅓ cup		
Baking soda	6 g	<¼ oz	1 teaspoon		
Salt	6 g	<¼ oz	1 teaspoon		

THE WET WORKS 2

INGREDIENT	Weight		Volume	Count	Prep
Unsalted butter	113 g	4 oz	8 tablespoons	1 stick	melted and cooled
Eggs	100 g	3½ oz		2 large	
Almond extract	6 g	¼ oz	1 teaspoon		

THE OPTIONS

INGREDIENT	Weight		Volume	Count	Prep
Nuts (walnuts, pecans, or almonds)			1 cup		chopped

THE EXTRAS

Baker's Joy or AB's Kustom Kitchen lube for the pan

✦ ✦ ✦ ✦ ✦

✦ Place an oven rack in position C and preheat the oven to 350°F.

✦ Prep a loaf pan (see pages 180–183) and set aside.

✦ Place the bananas and sugar in a large mixing bowl and mash them together with a potato masher until smooth (obviously this is impossible with all but the most overripe of bananas).

✦ Proceed with the **MUFFIN METHOD**. Combine the Wet Works 1 and then the Wet Works 2 in separate bowls, then combine together before adding the Wet Works to the Dry Goods.

✦ Fold in the nuts, if using, then pour the batter into the prepared loaf pan and bake for 50 minutes to 1 hour, until the temperature measures 210°F, or a toothpick inserted in the center comes out more or less clean, but banana bread should come out of the oven a little moister than most.

✦ Allow the bread to cool in the pan for 15 minutes, then remove it from the pan and place on a rack to cool completely before slicing.

✦ Tightly wrapped, this bread will keep at room temperature for 5 days.

Yield: One 9-inch loaf

✦ ✦ ✦ ✦ ✦

Muffin Variation

If you'd like to make banana muffins, prep a standard 12-hole muffin tin and use a disher to fill the holes. Bake on rack C for 30 minutes, until their temp reaches 210°F, or a toothpick inserted into the bottom of a muffin comes out more or less clean. Remove the muffins immediately to a cooling rack.

You'll get 12 standard muffins.

Notes:

Hardware:

Digital scale

Dry measuring cups

Measuring spoons

Chef's knife

Cutting board

Food processor

3 large mixing bowls

Medium saucepan

Rubber or silicone spatula

9 x 5 x 2½-inch aluminum loaf pan

Clean plastic spritz bottle

Cooling rack

Plastic wrap

Aluminum foil

Notes:

Nutty As A...

This particular application evolved from a fruitcake I made on TV a few years ago. That one was darned good, but having tinkered with it for a couple of years, I'm confident in saying that this one's better. You'll notice from the procedure that this is actually a quick bread rather than a cake and as such can be made into muffins no problem.

THE WET WORKS

INGREDIENT	Weight		Volume	Count	Prep
Golden raisins			¾ cup		
Dried cranberries			¾ cup		
Dried blueberries			¾ cup		
Dried cherries			¾ cup		
Dried apricots			½ cup		chopped
Candied ginger			¼ cup		chopped
Lemon zest				1 lemon	
Orange zest				1 orange	
Freshly ground cloves				5 whole	
Freshly ground allspice berries				3 berries	
Dark rum	227 g	8 oz	1 cup		
Unfiltered apple juice	170 g	6 oz	¾ cup		
Hard apple cider	113 g	4 oz	½ cup		
Unsalted butter	113 g	4 oz	8 tablespoons	1 stick	
Sugar	227 g	8 oz	1 cup		
Eggs	100 g	3½ oz		2 large	lightly beaten

THE DRY GOODS

INGREDIENT	Weight		Volume	Count	Prep
All-purpose flour	142 g	5 oz	1 cup		
Whole-wheat flour	142 g	5 oz	1 cup		
Baking powder	6 g	¼ oz	1 teaspoon		
Baking soda	6 g	¼ oz	1 teaspoon		

THE DRY GOODS (CONTINUED)

INGREDIENT	Weight		Volume	Count	Prep
Ground cinnamon	6 g	¼ oz	1 teaspoon		
Kosher salt	9 g	⅓ oz	1½ teaspoons		
Black pepper	3 g	⅛ oz	¼ teaspoon		

THE EXTRAS

INGREDIENT	Weight	Volume	Count	Prep
Pecans		½ cup		chopped, toasted
Brandy for spritzing				

✦ Mix together the raisins, dried fruits, ginger, lemon and orange zests, cloves, allspice, and rum, and soak overnight.

✦ The next day, stir in the apple juice, cider, butter, and sugar. Bring to a boil, stirring every couple of minutes. Reduce heat and simmer 10 minutes.

✦ Remove from heat and cool at least ½ hour. You can also keep this in the refrigerator, tightly covered, for up to a week. Just make sure you bring the batter to room temperature before proceeding. Stir in the eggs.

✦ Place an oven rack in position B and preheat the oven to 325°F. Prep a loaf pan (see pages 180–183) and set aside.

✦ Assemble the Wet Works and the Dry Goods by the **MUFFIN METHOD.** Pour the batter into the loaf pan and bake for 1 hour. Check for doneness with a skewer poked to the middle of the cake. If it comes out clean, you're done. If it comes out gooey, bake another 10 minutes, then check again.

✦ Set pan on cooling rack and spritz the cake with brandy. Cool completely before turning cake out of pan. Spritz with brandy again, then wrap tightly with plastic wrap and then foil.

✦ Although you can serve it right away, the loaf will improve with aging (or mellowing). Unwrap and spritz with brandy every second or third day for two weeks, then once a week for the next two weeks. After a month, the loaf will be ready to serve. Is it worth all that waiting? Yes.

Yield: One 9-inch loaf

Notes:

Hardware:

Digital scale

Dry measuring cups

Measuring spoons

Chef's knife

Cutting board

10-inch cast-iron skillet

Wooden spoon

Food processor

2 large mixing bowls

Whisk

Rubber or silicone spatula

Instant-read thermometer or toothpick

Oven mitt

Serving platter

Yes, it really has to be a cast-iron skillet. If you don't have one, get one—nothing else will do the job.

Notes:

Pineapple Upside-Down Cake

This recipe comes from my dad's mom, and it's the best I've ever sampled. It follows a procedure often found in old Southern fruit-based dessert recipes. The butter and sugar are melted in a pan before the fruit and batter are added. The transition from cake bottom to candy-like top is sublime. I suspect this procedure evolved out of pan-baked corn breads that usually call for melting fat in the skillet before pouring in the batter.

THE UPSIDE-DOWN PART

INGREDIENT	Weight		Volume	Count	Prep
Unsalted butter or margarine	113 g	4 oz	8 tablespoons	1 stick	
Dark brown sugar	227 g	8 oz	1 cup		
Canned pineapple in heavy syrup				6 slices	drained
Maraschino cherries				6	
Pecans	21 g	¾ oz	¼ cup		chopped
Pineapple juice	43 g	1½ oz	3 tablespoons		

THE DRY GOODS

INGREDIENT	Weight		Volume	Count	Prep
All-purpose flour	135 g	4¾ oz	1 cup		
Baking powder	3 g	⅛ oz	1 teaspoon		
Salt	2 g	<⅛ oz	¼ teaspoon		

THE WET WORKS

INGREDIENT	Weight		Volume	Count	Prep
Eggs	150 g	5¼ oz		3 large	
Pineapple juice	71 g	2½ oz	5 tablespoons		
Granulated sugar	198 g	7 oz	1 cup		

✦ ✦ ✦ ✦ ✦

✦ Place an oven rack in position C and preheat the oven to 350°F.

✦ Prepare the Upside-Down Part: Place a cast-iron skillet over medium heat and melt the butter. Add the dark brown sugar and stir until it dissolves, about 5 minutes. Remove the skillet from the heat and carefully place 1 pineapple slice in the center of the pan. Circle the other slices around the center slice. Put a cherry in the center of each pineapple slice and sprinkle the nuts evenly over the fruit. Top with the pineapple juice.

✦ Assemble the Dry Goods and the Wet Works via the **MUFFIN METHOD.**

✦ Pour the batter over the fruit in the skillet and bake for 40 minutes, until the temp of the cake reaches 210°F, or a toothpick inserted comes out clean. Put that oven mitt on your hand, remove the skillet from the oven, and allow to cool slightly.

✦ Set a platter on top of the skillet and carefully invert the cake.

✦ Consume mass quantities, and if there's any left, it will keep—tightly wrapped at room temperature—for 5 days.

Yield: One 10-inch hunk of tropical heaven, serves 6 to 8

> Yes, the nasty kind from a jar—but just this once.

1 Thermometer

2 Plate

3

4

THE WINNING RECIPE

✦ ✦ ✦

Recipe contests have added a number of classic recipes to the American zeitgeist. The Tunnel of Fudge Bundt Cake is one; Pineapple Upside-Down Cake is another.

Although the upside-down baking technique was in existence long before, the use of pineapple can be traced to Hawaiian plantation founder Jim Dole's introduction of canned pineapple in 1903. The Hawaiian Pineapple Company (now Dole Foods) ran an advertising campaign that featured pineapple recipes in all the popular women's magazines, turning a once-exotic fruit into a household staple. The 1925 campaign featured a recipe contest to be judged by representatives of the Fannie Farmer School and *Good Housekeeping* and *McCall's* magazines. The 100 best recipes would be published in a book the following year. More than 60,000 recipes were entered into the competition—of those, 2,500 were recipes for upside-down cake. The 1926 ad campaign actually used the number of upside-down cake recipes entered as an example of the dessert's popularity.

The original winning recipe featured either crushed pineapple or pineapple rings, and the maraschino cherries were added after the cake was baked.

Hardware:

Electric griddle or nonstick frying pan

Digital scale

Dry measuring cups

Wet measuring cups

Measuring spoons

Food processor

2 large mixing bowls

Whisk

Rubber or silicone spatula

Paper towel

2½-ounce disher

Pancake turner

Clean kitchen towel

Electric heating pad

Notes:

Buttermilk Pancakes

Yes, Virginia, pancakes are muffins—flat, griddle-cooked muffins. This means that next to waffles, their closest relatives are English Muffins, which, oddly enough, are not assembled via the Muffin Method.

THE DRY GOODS

INGREDIENT	Weight		Volume	Count	Prep
All-purpose flour	270 g	9½ oz	2 cups		
Baking powder	3 g	⅛ oz	1 teaspoon		
Baking soda	3 g	⅛ oz	½ teaspoon		
Salt	6 g	¼ oz	1 teaspoon		
Sugar	42 g	1½ oz	3 tablespoons		

THE WET WORKS

INGREDIENT	Weight		Volume	Count	Prep
Eggs	100 g	3½ oz		2 large	
Buttermilk	454 g	16 oz	2 cups		at room temperature
Unsalted butter	57 g	2 oz	4 tablespoons	½ stick	melted and slightly cooled

THE EXTRAS

Unsalted butter for the griddle

✦ ✦ ✦ ✦ ✦

✦ Heat an electric griddle to 350°F or place a nonstick frying pan over medium-low heat.

✦ Assemble the Dry Goods and the Wet Works via the **MUFFIN METHOD.**

✦ Set the batter aside to rest for 5 minutes.

✦ Test the griddle by flicking water on it. If the water dances across the griddle you're good to go.

✦ Rub the griddle down with a little butter, then wipe it up with a paper towel (there will be enough fat left to make a difference, believe me).

✦ Use the disher to ladle 1 scoop of batter onto the griddle and cook until bubbles form and the bottom is golden, approximately 3 minutes. Flip and cook until the second side is golden, another 2 minutes or so. Adjust the heat as necessary as you go along.

✦ Serve right away or keep the pancakes warm, layered inside the folds of a kitchen towel placed under a heating pad set to high. Leftover pancakes can be frozen, each separated by a sheet of waxed paper in zip-top bags, for up to 1 month. To reheat, brush them with butter and place them in the oven for 10 minutes at 350°F, or heck…do what I do and put them in the toaster.

Yield: Twelve 4-inch pancakes

Notes:

Why? To allow the leavening to create a bunch of little bubbles that will then blow up when the griddle heat is applied. These bubbles, along with the hydrating flour, will thicken the batter, making for cleaner, more orderly rounds.

Or you can just shoot the griddle with your infrared thermometer.

Hardware:

Electric or stove-top griddle or non-stick frying pan

Digital scale

Dry measuring cups

Wet measuring cups

Measuring spoons

Food processor

2 large mixing bowls

Whisk

Rubber or silicone spatula

Paper towel

2½-ounce disher

Pancake turner

Clean kitchen towel

Electric heating pad

Notes:

Whole Wheat Pancakes

Good and good for you. To the best of my knowledge, this is the only recipe I have that calls only for whole wheat flour. Most "whole wheat" recipes include at least some white flour because the oat and bran make gluten construction difficult. Luckily, we don't need no stinkin' gluten here.

THE DRY GOODS

INGREDIENT	Weight		Volume	Count	Prep
Whole wheat flour	284 g	10 oz	2 cups		
Baking powder	3 g	⅛ oz	1 teaspoon		
Baking soda	3 g	⅛ oz	½ teaspoon		
Salt	6 g	¼ oz	1 teaspoon		
Sugar	42 g	1½ oz	3 tablespoons		

THE WET WORKS

INGREDIENT	Weight		Volume	Count	Prep
Eggs	100 g	3½ oz		2 large	
Buttermilk	454 g	16 oz	2 cups		at room temperature
Unsalted butter	57 g	2 oz	4 tablespoons	½ stick	melted and slightly cooled

THE EXTRAS

Unsalted butter for the griddle

✦ ✦ ✦ ✦ ✦

✦ Heat an electric griddle to 350°F or place a stove-top griddle or nonstick frying pan over medium-low heat.

✦ Assemble the Dry Goods and the Wet Works via the **MUFFIN METHOD.**

✦ Set the batter aside to rest for 5 minutes.

✦ Test the griddle by flicking water on it. If the water dances across the surface you're good to go.

✦ Rub the griddle down with a little butter, then wipe it up with a paper towel.

✦ Use the disher to ladle 1 scoop of batter onto the griddle and cook until bubbles form in the batter and the bottom is golden, approximately 3 minutes. Flip and cook until the second side is golden, another 2 minutes or so. Adjust the heat as necessary as you go along.

✦ Serve right away or keep the pancakes warm, layered inside the folds of a kitchen towel placed under a heating pad set to high. Leftover pancakes (ha, that's funny) can be frozen, each separated by a sheet of waxed paper, in zip-top bags for up to 1 month. To reheat, brush them with butter and place them in the oven for 10 minutes at 350°F, or go ahead and put them in the toaster.

Yield: Twelve 4-inch pancakes

HOT OFF THE GRIDDLE

✦ ✦ ✦

If you want to make pancakes, do yourself a favor and ditch the pan. I've got a heavy aluminum nonstick griddle with a heat-resistant handle that can manage four cakes at a time. If I need to boost production, I go to an electric model (mine's made by Toastmaster). The point of a griddle, of course, is that there are no sides to get in the way of a spatula.

Hardware:

Waffle iron

Digital scale

Dry measuring cups

Wet measuring cups

Measuring spoons

Food processor

2 large mixing bowls

Whisk

Rubber or silicone spatula

2½-ounce disher

Clean kitchen towel

Electric heating pad

Notes:

Good name, huh? It would be even better if this waffle was of German origin. It's not . . . sorry.

Luft-Waffle

Thank goodness for the Dutch. Ever since they won their independence from Spain, the Dutch have been a progressive folk. That's why in 1611 (or thereabouts) English separatists seeking religious freedom smuggled themselves to Amsterdam. A decade later they headed to America. We call these folks the Pilgrims, and were it not for their sojourn in Holland, odds are we wouldn't enjoy as many of the Dutch creations that we do…like waffles. Of course I doubt even the progressive Dutch ever got around to serving them under fried chicken.

THE DRY GOODS

INGREDIENT	Weight		Volume	Count	Prep
All-purpose flour	270 g	9½ oz	2 cups		
Baking powder	3 g	⅛ oz	1 teaspoon		
Baking soda	3 g	⅛ oz	½ teaspoon		
Salt	6 g	¼ oz	1 teaspoon		

THE WET WORKS

INGREDIENT	Weight		Volume	Count	Prep
Unsalted butter	57 g	2 oz	4 tablespoons	½ stick	melted and slightly cooled
Eggs	150 g	5¼ oz		3 large	beaten
Buttermilk	454 g	16 oz	2 cups		at room temperature
Sugar	42 g	1½ oz	3 tablespoons		

THE EXTRAS

Baker's Joy or AB's Kustom Kitchen Lube for the waffle iron ✦ Unsalted butter and maple syrup for serving

✦　✦　✦　✦　✦

✦ Coat the waffle iron with the Baker's Joy and preheat it.

✦ Assemble the Dry Goods and the Wet Works via the **MUFFIN METHOD,** whisking the melted butter with the eggs before mixing in the remainder of the Wet Works.

✦ Set the batter aside to rest for 5 minutes. (Or refrigerate it for up to an hour.)

✦ Use the disher to ladle 2 full scoops of batter onto the waffle iron and spread the batter lightly with the back of the disher.

✦ Close the waffle iron top and bake until the waffle is golden and crisp on both sides and can be easily removed from the iron, about 3 to 4 minutes.

✦ Repeat until you've used up the batter.

✦ Serve immediately or keep warm in a 200°F oven or by swaddling in a clean kitchen towel and nuzzling under a heating pad set on high until ready to serve.

✦ Eat with plenty of butter and real maple syrup.

Yield: Six 8-inch round waffles

✦　✦　✦　✦　✦

Chocolate Variation

Add ½ cup cocoa powder to the Dry Goods and 1 teaspoon vanilla extract to the Wet Works, plus ½ cup of chocolate chips and you have chocolate waffles that can be topped with ice cream or whipped cream—I like them better than chocolate cake.

WAFFLE WARES

✦　✦　✦

Although I'm a longtime Villaware fan, I recently purchased one of the new "waffle bakers" from KitchenAid's Pro-Line series. It looks just like a professional two-sided "flip"-style iron and works just like one too. Now I can make two Belgian waffles at a time.

Hardware:

Waffle iron

Digital scale

Dry measuring cups

Wet measuring cups

Measuring spoons

Food processor

2 large mixing bowls

Whisk

Rubber or silicone spatula

2½-ounce disher

Clean kitchen towel

Electric heating pad

Small saucepan

Notes:

Piña Colada Waffles

A refreshing summer waffle—and a fine excuse to serve rum for breakfast.

THE DRY GOODS

INGREDIENT	Weight		Volume	Count	Prep
All-purpose flour	270 g	9½ oz	2 cups		
Baking powder	3 g	⅛ oz	1 teaspoon		
Baking soda	3 g	⅛ oz	½ teaspoon		
Salt	6 g	<¼ oz	1 teaspoon		

THE WET WORKS

INGREDIENT	Weight		Volume	Count	Prep
Eggs	150 g	5¼ oz		3 large	beaten
Unsalted butter	57 g	2 oz	4 tablespoons	½ stick	melted and slightly cooled
Coconut milk	454 g	16 oz	2 cups		at room temperature
Sugar	42 g	1½ oz	3 tablespoons		
Crushed pineapple	78 g	2¾ oz	½ cup		drained

THE EXTRAS

Baker's Joy or AB's Kustom Kitchen Lube for the waffle iron
Unsalted butter for serving ✦ Rum Sauce for serving (recipe opposite)

✦ ✦ ✦ ✦ ✦

✦ Coat the waffle iron with the Baker's Joy and preheat it.

✦ Assemble the Dry Goods and the Wet Works via the **MUFFIN METHOD,** whisking the eggs with the melted butter before mixing in the remainder of the Wet Works.

✦ Set the batter aside to rest for 5 minutes. (Or refrigerate it for up to an hour.)

✦ Use the disher to ladle 2 full scoops of batter onto the waffle iron and spread the batter lightly with the back of the disher.

✦ Close the waffle iron top and bake until the waffle is golden and crisp on both sides and can be easily removed from the iron, 5 to 6 minutes.

✦ Repeat until you've used up the batter.

✦ Serve immediately with Rum Sauce or keep warm in a 200°F oven or by swaddling in a kitchen towel and nuzzling under a heating pad set on high until ready to serve.

Yield: Six 8-inch round waffles

✦ ✦ ✦ ✦ ✦

Rum Sauce

INGREDIENT	Weight		Volume	Count	Prep
Unsalted butter	113 g	4 oz	8 tablespoons	1 stick	
Dark brown sugar	170 g	6 oz	¾ cup		
Dark rum	76 g	2⅔ oz	⅓ cup		
Vanilla extract	28 g	1 oz	½ teaspoon		
Salt			pinch		

✦ Melt the butter in a small saucepan over medium heat.

✦ Add the sugar and rum and whisk until the sugar is dissolved. Cook, whisking occasionally, until the sauce has thickened, 5 to 7 minutes.

✦ Add the vanilla and the salt and cook an additional 2 to 3 minutes.

✦ Serve over the Piña Colada Waffles.

or drink on the rocks . . . or throw into a blender with some ice and enjoy with a couple of the waffles.

Notes:

Hardware:

10-inch cast-iron skillet

Digital scale

Dry measuring cups

Wet measuring cups

Measuring spoons

Food processor

2 large mixing bowls

Whisk

Oven mitt or pot holder

Rubber or silicone spatula

Pancake turner

Serving plate

Notes:

Dutch Baby Bunnies

What the heck's a Dutch Bunny? Well it's like a cross between a popover and a pancake. And it's a great breakfast/brunch item.

THE DRY GOODS

INGREDIENT	Weight		Volume	Count	Prep
All-purpose flour	67 g	2⅜ oz	½ cup		
Vanilla sugar	28 g	1 oz	2 tablespoons		see note
Salt	2 g	<⅛ oz	¼ teaspoon		

THE WET WORKS

INGREDIENT	Weight		Volume	Count	Prep
Unsalted butter	57 g	2 oz	4 tablespoons	½ stick	
Milk	113 g	4 oz	½ cup		
Eggs	100 g	3½ oz		2 large	beaten

THE EXTRAS

INGREDIENT	Weight	Volume	Count	Prep
Apple wood–smoked bacon			3 strips	cooked and crumbled
Vanilla sugar for sprinkling				see Note
Pure maple syrup, fresh squeezed lemon juice, or powdered sugar for serving				

✦ ✦ ✦ ✦ ✦

Notes:

✦ Adjust an oven rack to position C and preheat oven to 400°F.

✦ Put the butter in the skillet and then place the skillet into the oven to melt the butter. Watch out—don't let it burn.

✦ Assemble the Dry Goods, the remaining Wet Works, and the bacon via the **MUFFIN METHOD.**

✦ Grab that pot holder and remove the skillet from the oven. Pour the batter into the skillet and bake for 20 minutes, or until golden brown.

✦ Remove the skillet from the oven and transfer the bunny to a serving plate.

✦ Sprinkle with more vanilla sugar and top with desired topping. Cut into wedges and serve immediately.

Yield: 2 servings

Note: To make vanilla sugar, bury three vanilla beans in a one-pound bag of granulated sugar. Store in an air-tight container for one week to allow the flavor (and aroma) of the vanilla to permeate the sugar.

Cheesy Poof

This one is dedicated to a little boy named Cartman in South Park, Colorado. I know he'd really prefer to have a chocolate chicken pot pie on the side, but that will have to wait for another book.

Hardware:

9 x 5 x 2½-inch loaf pan

Digital scale

Dry measuring cups

Wet measuring cups

Measuring spoons

Food processor

2 large mixing bowls

Whisk

Rubber or silicone spatula

Instant-read thermometer or toothpick

Cooling rack

Notes:

THE DRY GOODS

INGREDIENT	Weight		Volume	Count	Prep
All-purpose flour	270 g	9½ oz	2 cups		
Baking powder	7 g	¼ oz	2 teaspoons		
Dry mustard	9 g	⅓ oz	1½ teaspoons		

THE WET WORKS

INGREDIENT	Weight		Volume	Count	Prep
Eggs	100 g	3½ oz		2 large	
Vegetable oil	43 g	1½ oz	3 tablespoons		
Milk	227 g	8 oz	1 cup		
Granulated sugar	15 g	½ oz	1 tablespoon		
Kosher salt	6 g	<¼ oz	1 teaspoon		

THE EXTRAS

INGREDIENT	Weight		Volume	Count	Prep
Cheddar cheese	227 g	8 oz	2 cups		shredded
Baker's Joy or AB's Kustom Kitchen Lube and parchment for the pan					

I like sharp New York cheddar for this. Vermont-style (white) cheddar would be okay too, but some part of me would miss the color.

✦ ✦ ✦ ✦ ✦

✦ Place an oven rack in the C position and preheat the oven to 375°F.

✦ Prep the loaf pan (see pages 180–183) and set it aside.

✦ Assemble the Dry Goods and the Wet Works via the **MUFFIN METHOD,** tossing the cheddar cheese with the Dry Goods after they've gone for their spin.

✦ Pour the batter into the loaf pan and bake for 40 minutes, until the internal temperature is 210°F, or a toothpick inserted in the center comes out clean.

✦ Remove the loaf from the pan immediately and allow to cool completely on a rack.

Yield: 1 loaf

Note: You can pan-fry slices of this in some butter for an all-in-one grilled cheese. Now that's some serious love, children.

Notes:

Herb Loaf

Although it goes together like a muffin, I don't think this bread is very breakfasty. It is, however, a nice luncheon or dinner bread. You can also make it as muffins (see the variation, opposite).

Hardware:

Digital scale

Dry measuring cups

Wet measuring cups

Measuring spoons

Chef's knife

Cutting board

9 x 5 x 2½-inch loaf pan

Food processor

2 large mixing bowls

Whisk

Rubber or silicone spatula

Instant-read thermometer or toothpick

Cooling rack

Notes:

THE DRY GOODS

INGREDIENT	Weight		Volume	Count	Prep
All-purpose flour	270 g	9½ oz	2 cups		
Whole wheat flour	284 g	10 oz	2 cups		
Baking powder	7 g	¼ oz	2 teaspoons		
Salt	12 g	½ oz	2 teaspoons		
INCLUDE THE FOLLOWING:					
Fresh chives			¼ cup		chopped
Fresh dill			¼ cup		chopped
Caraway seeds			2 teaspoons		
OR					
Fresh rosemary			2 tablespoons		chopped
Fresh thyme			2 tablespoons		chopped
Freshly ground black pepper	12 g	½ oz	2 teaspoons		

THE WET WORKS

INGREDIENT	Weight		Volume	Count	Prep
Sugar	56 g	7½ oz			
Olive oil	213 g	7 oz	1 cup		
Eggs	200 g	7 oz		4 large	
Egg yolks	40 g	1⅓ oz		2 large	
Buttermilk	284 g	10 oz	1¼ cups		

THE EXTRAS

Baker's Joy or AB's Kustom Kitchen Lube and parchment for the pan

✦   ✦   ✦   ✦   ✦

✦ Place an oven rack in position C and preheat the oven to 350°F.

✦ Prep a loaf pan (see pages 180–183) and set aside.

✦ Assemble the Dry Goods and the Wet Works via the **MUFFIN METHOD,** mixing the herbs and spices in with the Dry Goods after they've gone for their spin.

✦ Pour the batter into the prepared loaf pan and bake for 45 to 50 minutes, until the temperature reaches 210°F, or a toothpick inserted in the center of the loaf comes out clean.

✦ Cool in the pan for 15 minutes, then remove to a rack to cool completely.

✦ Wrapped tightly and stored at room temperature, the loaf will keep for up to a week.

Yield: One 9-inch loaf

✦   ✦   ✦   ✦   ✦

Muffin Variation

If you'd like to make herb muffins, the only difference in assembly is that you'll fold the herbs in at the end rather than mixing them with the dry ingredients. Spray a standard 12-hole muffin tin with Baker's Joy, and use a 2½-ounce disher to fill the holes. Bake in a 350°F oven on a rack placed in the C position for 35 minutes, until the temperature reaches 210°F, or a toothpick inserted into the center of a muffin comes out more or less clean. Remove the muffins from the pan to a cooling rack immediately.

You'll get 12 standard muffins.

Notes:

Hardware:

10-inch cast-iron skillet

Small saucepan

Digital scale

Dry measuring cups

Wet measuring cups

Measuring spoons

Food processor

2 large bowls

Whisk

Rubber or silicone spatula

Oven mitt or pot holder

Instant-read thermometer
or toothpick

Serving plate

Cutting board

Bread knife

Notes:

Ground from approximately 2 cups of stone-ground yellow grits or coarse polenta. (Bob's Red Mill does the job nicely. I also like Ansen Mills cornmeal, although it's tough to find outside the South.) Simply spin in a food processor until the pieces break down to roughly half their size. Do not use commercial cornmeal.

Cornbread No Chaser

As straightforward—and good—as it gets. The key is the cast-iron skillet. Accept no substitute.

THE WET WORKS 1

INGREDIENT	Weight		Volume	Count	Prep
Cornmeal	203 g	7¼ oz	1½ cups		
Milk	284 g	10 oz	1¼ cups		heated

THE DRY GOODS

INGREDIENT	Weight		Volume	Count	Prep
All-purpose flour	135 g	4¾ oz	1 cup		
Baking powder	9 g	<½ oz	1 tablespoon		
Kosher salt	6 g	¼ oz	1½ teaspoons		
OR					
Table salt	6 g	¼ oz	1 teaspoon		

THE WET WORKS 2

INGREDIENT	Weight		Volume	Count	Prep
Vegetable oil	106 g	3¾ oz	½ cup		
Eggs	100 g	3½ oz		2 large	

THE EXTRAS

INGREDIENT	Weight		Volume	Count	Prep
Unsalted butter for the skillet, plus additional for serving	14 g	½ oz	1 tablespoon		

If you can get a tangy, European-style cultured butter, you should.

✦ ✦ ✦ ✦ ✦

✦ Place the skillet on an oven rack at position B and preheat the oven to 450°F.

✦ In a large bowl soak the cornmeal in the milk for 15 minutes.

✦ Assemble the Dry Goods and the Wet Works via the **MUFFIN METHOD,** beating the oil and eggs into the cornmeal-milk mixture before combining with the Dry Goods.

✦ Put an oven mitt over your hand and remove the hot skillet from the oven. Lube the skillet with the butter, pour in the batter, and bake about 25 minutes, or until the interior hits 190°F and the cornbread is golden brown and delicious looking.

✦ Invert a serving plate over the top of the skillet and hold it with one hand while flipping the pan with the other. The cornbread should come right out of the pan. Now, slide it onto a cutting board, slice into wedges, and return to the plate.

✦ Brush with additional butter if you like…or not. Heck, it's cornbread—do what you want. Me, I eat it with collard greens and black-eyed peas…yum.

✦ And if you don't eat it all in one sitting, this cornbread will keep, tightly wrapped at room temperature, for 5 days.

Yield: 6 servings

KNOW YOUR CORNMEAL

✦ ✦ ✦

Cornmeal is a funny thing, or at least the way it's packaged is funny. Here's a quick guide:

✦ **GRITS.** Usually ground from hominy, that is corn that has been treated with alkali to remove the germ and kernel coat. Hominy grits are almost always white.

✦ **POLENTA.** Essentially coarse yellow cornmeal. The grind is a little coarser than grits.

✦ **MEDIUM CORNMEAL.** Finer than either of the above. Can be used in this recipe if you must.

✦ **WHOLE-GRAIN CORNMEAL.** Contains most of the germ and kernel coat. I'm not a huge fan simply because the kernel coat never really breaks down. If you like your cornbread to have backbone and bite, go for this one.

Hardware:

Deep fryer

Digital scale

Dry measuring cups

Wet measuring cups

Measuring spoons

Box grater

Chef's knife

Food processor

2 large mixing bowls

Whisk

Rubber or silicone spatula

1-ounce disher

Tongs

Newspaper or paper towels

Cooling rack

Notes:

If you don't have an electric deep fryer, so be it. Use a heavy Dutch oven and a fry thermometer. And for goodness sake, be careful, will ya?

Jalapeño Hush Puppies

Although this small fritter has many alleged histories, only one rings true for me. Back in the South a few zillion years ago, men did a lot of hunting…hunting with hounds. At the end of the day, some of whatever got shot or caught was fried up in a pan—usually with a cornmeal crust. The hounds, driven to distraction by the smells, whimpered and cried. To silence the din, the cook would fry up some of the cornmeal batter and throw it to the collected cur. Thus, the puppies were hushed. By adding some heat to these, I've played a rather mean trick on my own hound, Matilda, though she doesn't really seem to notice or mind.

THE DRY GOODS

INGREDIENT	Weight		Volume	Count	Prep
All-purpose flour	150 g	$5\frac{1}{3}$ oz	1 cup plus 2 tablespoons		
Medium-grind cornmeal	142 g	5 oz	1 cup plus 3 tablespoons		
Baking powder	3 g	$\frac{1}{8}$ oz	1 teaspoon		
Baking soda	1.5 g	$<\frac{1}{8}$ oz	$\frac{1}{4}$ teaspoon		
Cayenne pepper	2 g	$\frac{1}{8}$ oz	$\frac{1}{2}$ teaspoon		
Salt	12 g	$\frac{1}{2}$ oz	2 teaspoons		

THE WET WORKS

INGREDIENT	Weight		Volume	Count	Prep
Buttermilk	340 g	12 oz	$1\frac{1}{2}$ cups		
Creamed corn	8.5 ounce can				
Onion			$\frac{1}{3}$ cup		finely grated
Jalapeño chile			3 tablespoons	about $1\frac{1}{2}$ chiles	seeded and minced

THE EXTRAS

INGREDIENT	Weight		Volume	Count	Prep
Peanut oil for frying	3.63 kg	128 oz	4 quarts		

Notes:

✦ ✦ ✦ ✦ ✦

✦ Pour the oil into a deep fryer (mine usually stays loaded) and set it for 375°F.

✦ Assemble the Dry Goods and the Wet Works via the **MUFFIN METHOD.** When mixing, keep in mind there will be lumps...that's life.

✦ Let the batter rest at room temperature for 10 minutes.

✦ Using the disher, very carefully spoon the batter into the oil. Do not crowd the fryer—small batches of 4 to 5 puppies at a time work best, depending on the size of your fryer. Once the fritters float to the surface, flip them over if you can (they won't want to). Remove with tongs when golden brown, 3 to 3½ minutes total, and drain on a rack set over newspaper or paper towels.

Yield: About 36 puppies

I use peanut oil because I live in Georgia and we've got plenty to go around. I also think it smells the house up less than regular vegetable oil—and if you strain it through a fine-mesh sieve, I find it can be re-used more often than most other oils.

Hardware:

Medium mixing bowl

Whisk

Rubber or silicone spatula

Dough cutter

Clean kitchen towel

Rolling pin or pasta machine

Tailor's ruler

1 or 2 jellyroll or half sheet pans

Pancake turner

Cooling rack

Notes:

Seedy Crisps

I love crackers, and I'm not about to trust their production to anyone (especially elves in hollow trees). These whole wheat crackers are tasty and toothsome and way better than store bought—and they're easy to make. So no excuses—get cracking!

THE DRY GOODS

INGREDIENT	Weight		Volume	Count	Prep
Whole wheat flour	142 g	5 oz	1 cup		
All-purpose flour	135 g	4¾ oz	1 cup		
Salt	9 g	⅓ oz	1½ teaspoons		
Baking powder	8 g	⅓ oz	1½ teaspoons		
Sesame and poppy seeds			⅔ cup mixed		

THE WET WORKS 1

INGREDIENT	Weight		Volume	Count	Prep
Olive oil	40 g	1½ oz	3 tablespoons		

THE WET WORKS 2

INGREDIENT	Weight		Volume	Count	Prep
Water	227 g	8 oz	1 cup		

THE EXTRAS

Flour for rolling ✦ Parchment for the pan

✦   ✦   ✦   ✦   ✦

✦ Place oven racks at positions B and C and preheat the oven to 450°F.

✦ In a medium bowl, prepare the Dry Goods according to the **MUFFIN METHOD**. Add the oil and stir until just combined. Then add water slowly until the dough just comes together (you might not use all the water).

✦ Turn the dough out onto a floured surface and knead 3 times. Divide the dough into 8 pieces, cover with a kitchen towel, and allow to rest for 15 minutes.

✦ Roll out the dough according to the directions below and place it on a parchment-lined baking sheet.

✦ Bake according to directions below.

✦ Remove crisps from oven and place on a cooling rack. Break into pieces of desired size.

For a thin snacking cracker:

✦ Divide the dough into fourths, roll out as thin as the seeds will allow (about $1/16$ inch).

✦ Bake for 4 minutes, flip, then bake for 2 to 3 more minutes, or until toasty brown.

For a thicker dipping cracker:

✦ Divide the dough in half, roll out to $1/8$ inch thick, and then stretch the dough out on 2 baking sheets. Try to wrap the dough edges over the pan to prevent shrinkage. As the crisps bake, the dough begins to pull away from the pan.

✦ Bake for 6 minutes, flip, then bake for another 4 to 5 minutes, or until toasty brown.

Yield: The amount will vary according to cracker size, but this recipe will yield about 24 2 x 2-inch crackers or a dozen 4 x 4-inch crackers.

Notes: Salt is not necessary on these crackers.
If using a pasta machine, do not use a setting thinner than 5.

Notes:

A Carrot Cake by Any Other Name

Judging a baked good by its moniker can sometimes lead you down a road of conflict and confusion. Names can mess with your mind for the simple reason that names often conceal the true nature of things. Take two of our culture's most beloved desserts: cheesecake and carrot cake. In my estimation, neither are cakes—and calling them such is like calling a grizzly bear a big dog. Sure, on some levels a bear is like a dog…but not on any of the levels that count. So, why aren't they cakes? After all, they're both round, they're baked in pans—heck, the carrot cake even has frosting and filling. Who's to say they're not cakes?

I believe classifications should shed light on a recipe, not shroud it with mystery. The truth is, a cheesecake is a custard pie, and carrot cake is a quick bread and therefore a muffin. Knowing this is important because it will give you a better sense of how things work, and if you understand how things work you can make your recipes work better—no question about it. Understanding that a carrot cake is really a big muffin means you don't have to do a lot of the things that so many carrot cakes ask you to do.

Here's the ingredient list from an old carrot cake recipe that's been bouncing around my family since at least the late 1940s.

THE DRY GOODS

INGREDIENT	Weight	Volume	Count	Prep
All-purpose flour	6 oz			sifted
Baking powder		$3/4$ rounded teaspoon		
Baking soda		$3/4$ teaspoon		
Ground cinnamon		$3/4$ rounded teaspoon		
Salt		$1/4$ rounded teaspoon		

THE WET WORKS

INGREDIENT	Weight	Volume	Count	Prep
Sunflower seed oil		just under 1 cup		
Powdered sugar	12 oz	$1/2$ cup		
Eggs	6 oz			
Raw carrot			1 large	grated

THE EXTRAS

Vegetable oil for greasing the pan ✦ Flour for dusting the pan

✦ ✦ ✦ ✦ ✦

Notes:

Notes:

First off, this is a crummy formula—not because the food it produces is bad (we'll get to that later), but because it's way too open to interpretation. Why? Rounded teaspoons for one. That's an annoyingly nebulous amount, if you ask me. And even if I knew how to "round" a teaspoon, how, pray tell, do you "round" three quarters of a teaspoon? Do you round a half teaspoon, and round a quarter teaspoon? Wouldn't that be…a teaspoon? Who knows? And why sunflower seed oil? And why "just under"…why? Did you save some out to lube the pan? Powdered sugar? Do you mean confectioners' sugar or sugar that's been taken for a spin around the food processor? Just reading this gets me in such a snit that I don't even want any stupid ol' cake anymore.

Still, I decided to attempt quantification and came up with this:

THE DRY GOODS

INGREDIENT	Weight		Volume	Count	Prep
All-purpose flour	170 g	6 oz	1$\frac{1}{8}$ cups		sifted
Baking powder	3 g	$\frac{1}{8}$ oz	1 teaspoon		
Baking soda	5 g	<$\frac{1}{4}$ oz	$\frac{3}{4}$ teaspoon		
Ground cinnamon	5 g	<$\frac{1}{4}$ oz	1 teaspoon		
Salt	2 g	<$\frac{1}{8}$ oz	$\frac{1}{4}$ rounded teaspoon		

THE WET WORKS

INGREDIENT	Weight		Volume	Count	Prep
Sunflower seed oil	213 g	7$\frac{1}{2}$ oz	$\frac{7}{8}$ cup		
Confectioners' sugar	340 g	12 oz			
Eggs	150 g	5$\frac{1}{4}$ oz		3 large	

THE EXTRAS

INGREDIENT	Weight		Volume	Count	Prep
Raw carrot	170 g	6 oz		1 large	grated
Nuts			2 cups		chopped

Vegetable oil for greasing the pan ✦ Flour for dusting the pan and parchment

✦   ✦   ✦   ✦   ✦

Here are the instructions for assembling the formula from the original recipe:

✦ Preheat the oven to 350°F.

✦ Sift the dry ingredients into a large bowl.

✦ Put the oil in a large bowl, add the sugar, and beat them together using an electric hand mixer. Add the eggs, one by one, beating well in between each addition. Fold in the dry ingredients, and lastly stir in the grated carrot and nuts.

✦ Line a 10-inch round cake pan with parchment and lightly grease and flour it. Shake out all excess flour.

✦ Spoon the mixture into the pan and bake for 45 minutes, then lower the heat to 325°F and bake for another 20 minutes. Test for doneness by pushing a knife into the center of the cake—if it comes out clean, the cake is done. If the knife has a trace of cake mixture sticking to its blade, put the cake back in the oven and test again after another 15 minutes.

✦ Remove the cake from the oven, allow it to cool for 5 minutes in the pan, then turn it out onto a cooling rack. Carefully peel off the paper and allow the cake to cool completely.

✦　✦　✦　✦　✦

I made it, and it was a disgusting oily, oversweet, dense mess that tasted of nothing remotely resembling carrots—which I always figured was the whole reason for carrot cake. Since I knew that deep down I was dealing with a muffin, I figured I could up the flour to give the cake enough structure to catch the rise. I also figured I could replace at least three quarters of that oil with an acidic liquid like buttermilk or, better yet, yogurt. Adding acid changes the crumb of baked goods; they brown better and retain moisture better, which leads to better flavor. I also figured that since neither creaming nor kneading were involved, I could do away with the mixer. Since I already do my sifting in a food processor, why not do it all in the food processor? All, that is, but the carrots—machine-grated carrots are always a sopping mess, so I stuck with a manual box grater, but I upped the amount to try to swing the flavor back where it needed to be. And I ditched those darned nuts. This wasn't fruitcake after all.

Notes:

Although it is a first cousin. If you don't believe me, check the recipe on pages 100–101.

18-Carrot Cake

So here is my next and what became the final version—not too dense, not too light. Plenty of flavor; firm but not wet; smooth, but not too smooth (because if a good carrot cake is really a muffin, it shouldn't have the even, refined texture of a wedding cake). I've called it 18-Carrot Cake.

Hardware:

9 x 3 spun-aluminum round pan

Digital scale

Dry measuring cups

Wet measuring cups

Measuring spoons

Box grater

Chef's knife

Cutting board

2 large mixing bowls

Food processor

Rubber or silicone spatula

Instant-read thermometer or toothpick

Cooling rack

Stand mixer with paddle attachment or electric hand mixer

Bread knife

Lazy Susan

> There aren't actually 18 carrots in this cake, only 8 ounces. I just thought the name sounded cool.

THE DRY GOODS

INGREDIENT	Weight		Volume	Count	Prep
All-purpose flour	241 g	8$\frac{1}{2}$ oz	1$\frac{3}{4}$ cups		
Baking powder	3 g	$\frac{1}{8}$ oz	1 teaspoon		
Baking soda	5 g	$\frac{1}{4}$ oz	$\frac{3}{4}$ teaspoon		
Ground cinnamon	5 g	$\frac{1}{4}$ oz	1 teaspoon		
Kosher salt	2 g	$\frac{1}{8}$ oz	$\frac{1}{2}$ teaspoon		

THE WET WORKS

INGREDIENT	Weight		Volume	Count	Prep
Whole vanilla yogurt			$\frac{3}{4}$ cup		
Vegetable oil	57 g	2 oz			
Granulated sugar	284 g	10 oz			
Eggs	150 g	5$\frac{1}{4}$ oz		3 large	

THE EXTRAS

INGREDIENT	Weight		Volume	Count	Prep
Raw carrot	227 g	8 oz		3 to 4	grated
Baker's Joy or AB's Kustom Kitchen Lube and parchment for the pan					

✦ ✦ ✦ ✦ ✦

Notes:

✦ I chose a 9 x 3-inch spun-aluminum pan and you should, too. Prep it and set it aside.

✦ Crank your oven to 350°F and move a rack to position C.

✦ Grate 3 to 4 carrots on the small holes of a box grater until you've got 8 ounces. Place the carrots in a large mixing bowl.

✦ Combine the Dry Goods according to the **MUFFIN METHOD**, then dump the Dry Goods on top of the carrots and stir.

✦ Add the Wet Works to the food processor bowl and spin until thoroughly combined and slightly thickened.

As the eggs emulsify the mixture, it will indeed thicken a bit.

✦ Combine the carrot mixture with the Wet Works according to the **MUFFIN METHOD**.

✦ Pour the batter into the prepared pan.

✦ Bake for 45 minutes, then lower the oven temperature to 325°F and bake for another 15 minutes.

✦ Temp the center of the cake. You're looking for 205°F to 210°F. If you don't have a thermometer (I'm sure you do—you probably just loaned it to your neighbor), you can check for doneness with a toothpick. If it comes out with batter clinging to it, bake another 5 minutes and check again (that'll give you time to go to your neighbor's and get your thermometer back).

✦ Remove the pan from the oven and allow the cake to cool 15 minutes. Although the refrigerator is too cool and would probably cause condensation on and around the cake, which would be a bad thing, setting the pan in a cool place—the garage, a screened-in porch, the proverbial windowsill—would be a good idea.

✦ Turn the cake out onto a rack to cool completely.

✦ While the cake cools, contemplate the frosting.

Here's what my old recipe for Cream Cheese Frosting called for:

INGREDIENT	Weight		Volume	Count	Prep
Unsalted butter	170 g	6 oz	12 tablespoons	1½ sticks	
Cream cheese	170 g	6 oz	12 tablespoons		
Vanilla extract	5 g	¼ oz	½ teaspoon		
Powdered sugar	227 g	8 oz			

Notice, please, the amazingly inconvenient amounts: a stick and a half of butter and three quarters of a block of cream cheese. And what about the skimpy yield? I don't know about you, but I want three layers and more than a millimeter of frosting in between each.

For better Cream Cheese Frosting, I decided to go with:

INGREDIENT	Weight		Volume	Count	Prep
Unsalted butter	113 g	4 oz	8 tablespoons	1 stick	cut in pieces and softened for 10 minutes
Cream cheese	454 g	16 oz		two 8-ounce bricks	
Vanilla extract	5 g	¼ oz	½ teaspoon		
Confectioners' sugar	255 g	9 oz			

✦ Rig up your electric mixer with the paddle attachment.

✦ Cream the butter and cream cheese together at 50 percent power until smooth. Add the vanilla and beat until it's integrated.

✦ Drop the speed to 25 percent and slowly add the sugar in 4 batches, beating until smooth between each addition.

✦ Refrigerate the frosting for 5 to 10 minutes before frosting the cake.

✦ Split and frost the cake as shown.

Yield: Enough frosting for 1 cake split into 4 layers

Although I often make my own "extra-fine" sugar in a food processor, in this case you really need to use XXX sugar, which is a good bit finer than anything you can manage with a home machine. If your sugar is old and lumpy, you'll want to either spin it for a few seconds in the food processor or sift it.

I know, you're tempted to use the food processor here, but don't. The texture will never be right.

Cutting a Cake into Layers

BAKING A CAKE IN A SINGLE DEEP PAN will make it easier to cut it into three or even four layers. I've tried a lot of cake-splitting methods, some of which were kinda wacky, and I've come to the conclusion that all you need is a bread knife and a lazy Susan.

Unless you're mighty good with a knife, you'll want to mark the layers so that you can re-align them into the proper orientation after frosting. So before you cut, slice a small notch out of the side of the cake. This will function as pilot marks later on.

The knife moves straight across while the cake is rotated counterclockwise. If you're left-handed, everything reverses. **Always cut from the bottom layers up.**

I fight the urge to saw the knife through the cake. Once you get the hang of it, you'll be able to cut the entire layer by the time the blade is halfway through the cake.

Cut from bottom up

Stopping point for knife

Notch for alignment

Hardware:

Deep fryer (or Dutch oven and fry thermometer)

Digital scale

Dry measuring cups

Wet measuring cups

Measuring spoons

1 large mixing bowl

Food processor

Electric stand mixer

Rubber or silicone spatula

Teaspoon

Tongs

Newspaper or paper towels

Cooling rack

Small bowl

Notes:

Ricotta Clouds

These are doughnuts without holes, and—although the viscosity of the Wet Works requires the batter be brought together with a stand mixer—in reality, this is another muffin (albeit a fried one).

THE DRY GOODS

INGREDIENT	Weight		Volume	Count	Prep
All-purpose flour	270 g	9½ oz	2 cups		
Baking powder	14 g	½ oz	4 teaspoons		
Kosher salt	4 g	⅛ oz	¾ teaspoon		

THE WET WORKS

INGREDIENT	Weight		Volume	Count	Prep
Eggs	250 g	8¾ oz		5 large	
Whole milk ricotta cheese	454 g	1 lb			
Sugar	70 g	2½ oz	5 tablespoons		
Pure vanilla extract	5 g	¼ oz	½ teaspoon		
Lemon extract	9 g	⅓ oz	1 teaspoon		

THE EXTRAS

INGREDIENT	Weight		Volume	Count	Prep
Sugar for coating	105 g	3¾ oz	½ cup		
Vegetable oil for frying					

Notes:

✦ Preheat the vegetable oil in a deep fryer to 360°F.

✦ Combine the Dry Goods in a large bowl, according to the **MUFFIN METHOD.**

✦ Place the eggs in the bowl of a stand mixer and beat. Add the ricotta, sugar, and vanilla and lemon extracts and beat until well combined. Proceed mixing the wet into the dry via the **MUFFIN METHOD.**

✦ One spoonful at a time, gently drop the dough into the hot oil and cook until golden, about 1 to 2 minutes per side. Cook only 3 to 5 at a time so as not to overcrowd the fryer.

✦ Remove to a cooling rack placed over newspaper and drain for 3 to 4 minutes.

✦ Pour the additional sugar into a small bowl.

✦ Roll the clouds in the sugar and serve immediately.

Yield: 50 to 60 clouds

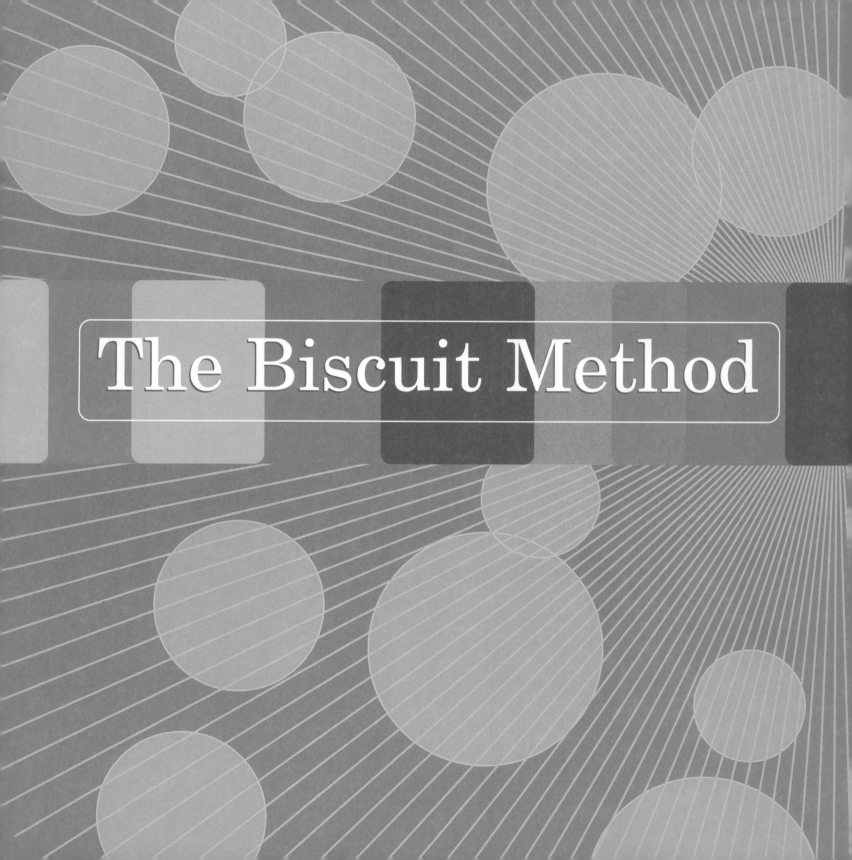

The Biscuit Method

The Biscuit Method refers to any procedure that calls for solid fat to be "cut" into flour. The two most common applications are biscuits (a baked good category that includes scones, shortbread, most cobblers, and more than a few dumplings) and pie crust, which is nothing but a biscuit with very little water and no leavening.

Your mission is to create a rollable, cuttable dough that bakes quickly into a tender structure that is finer than that produced by the Muffin Method but coarser than that produced by the Creaming Method. With the exception of pie crusts, all Biscuit Method products are raised with chemical leaveners and are therefore considered quick breads.

PHASE I BISCUIT

✦ ✦ ✦

INGREDIENTS:
2 cups White Lily All-Purpose Flour
1 tablespoon baking powder
1 teaspoon salt
¼ cup shortening
⅔ to ¾ cup milk or buttermilk

INSTRUCTIONS:
Preheat oven to 500°F. Lightly spray a baking sheet with non-stick cooking spray. Measure flour (spoon into measuring cup and level off) and place in bowl. Blend in baking powder and salt. Cut in shortening until mixture resembles coarse crumbs. With a fork, blend in just enough milk until dough leaves sides of bowl. (Too much milk makes dough sticky, while too little makes biscuits dry.) Knead gently 2 to 3 times on lightly floured surface. Roll dough to ½-inch thickness, cutting without twisting biscuit cutter. Place on baking sheet, 1-inch apart for crisp sided biscuits, almost touching for softer sided biscuits. Bake 8 to 10 minutes or until golden brown. Brush tops with melted butter if desired. Serve warm.

Makes twelve 2-inch biscuits.

Reprinted with the kind permission of the White Lily Foods Co., Inc.

Building a Better Biscuit

Biscuits are the first baked food, possibly the first food that I tried to figure out. I've baked more versions of the simple buttermilk biscuit than you can wag a butter knife at. To my mind buttermilk biscuits are vastly superior to either baking powder biscuits or Northern "cream" biscuits. Heck, the average Northerner knows even less about biscuits than the average Southerner does about pastrami…or Boston baked beans.

And yet every single one of them was assembled the same way. For years I lived and loved by my grandmother's biscuit recipe—the one on the back of the White Lily flour bag.

But after a few years, my shortening-dulled palate started craving a little more complexity. So I switched from shortening to lard. The pig fat upped the tenderness quotient and brought a subtle but noticeable flavor to the party—but not quite noticeable enough. I decided to ditch the buttermilk and go with yogurt.

This is what the ingredient list looked like:

THE DRY GOODS

INGREDIENT	Weight		Volume	Count	Prep
All-purpose flour	270 g	9½ oz	2 cups		
Baking powder	14 g	½ oz	4 teaspoons		
Baking soda	2 g	<⅛ oz	¼ teaspoon		
Salt	6 g	¼ oz	1 teaspoon		

THE FAT

INGREDIENT	Weight		Volume	Count	Prep
Unsalted butter	43 g	1½ oz	3 tablespoons		
Lard	14 g	½ oz	1 tablespoon		chilled
Plain yogurt			1 cup		full fat, if possible, or lowfat; not skim!

◆　◆　◆　◆　◆

This produced a chewier and tangy biscuit, but still a little dry. **In the next version** I upped the yogurt to 1¼ cups and the biscuits picked up some moistness. This was a very good biscuit indeed. But greedy little man that I am, I wanted more. I still thought the biscuit needed more beef (figuratively not literally), and I still wasn't happy with the color, so I decided on a radical change of course. I decided to go all butter for flavor and to use both buttermilk and yogurt, again for flavor and more moisture. An egg added structural "oomph" as well as richness. Why does a moist dough make a better biscuit? More water makes for more steam inside the dough and that means better leavening potential. It also means the dough can stay plastic longer, thus harnessing that leavening.

Notes:

Although any fat rendered from a hog can be sold as lard, the best is leaf lard, which comes from around the bowels and kidneys. Accept no substitute.

Non-fat yogurts almost always contain modified cornstarch. Mmm . . . none for me today, thanks.

Hardware:

Digital scale

Dry measuring cups

Wet measuring cups

Measuring spoons

Chef's knife

Cutting board

Food processor

Large mixing bowl

Box grater

Rubber or silicone spatula

Wax paper

Bench scraper

2-inch round biscuit cutter or pastry ring

Baking sheet

Kitchen-towel–lined basket

Phase III Biscuit

(My new favorite biscuit.)

THE DRY GOODS

INGREDIENT	Weight		Volume	Count	Prep
All-purpose flour	270 g	$9\frac{1}{2}$ oz	2 cups		
Baking powder	14 g	$\frac{1}{2}$ oz	4 teaspoons		
Baking soda	3 g	$\frac{1}{8}$ oz	$\frac{1}{2}$ teaspoon		
Salt	6 g	$<\frac{1}{4}$ oz	1 teaspoon		

THE FAT

INGREDIENT	Weight		Volume	Count	Prep
Unsalted butter	57 g	2 oz	4 tablespoons	$\frac{1}{2}$ stick	frozen solid

THE LIQUID

INGREDIENT	Weight		Volume	Count	Prep
Buttermilk	227 g	8 oz	1 cup		full fat if possible
Plain yogurt			$\frac{1}{3}$ cup		full fat, if possible, or lowfat; no skim
Eggs	50 g	$1\frac{3}{4}$ oz		1 large	

THE EXTRAS

Flour for rolling out the dough

✦   ✦   ✦   ✦   ✦

✦ Place an oven rack in position C and preheat your oven to 450°F.

✦ Take the Dry Goods for a spin in the food processor, then dump them into a large bowl. Place a box grater on top of the bowl.

✦ Remove the butter from the freezer and grate it through the big holes of the box grater into the Dry Goods. Toss the butter flakes into the flour mixture—just slide your fingers deep into the bowl and toss over and over to evenly distribute them.

✦ Using only the tips of your fingers, rub the butter into the Dry Goods. Some people do this with one of those ridiculous pastry blenders that looks like a curved potato masher. This will do exactly one third of the work in exactly twice the time. Don't use the food processor either—save it for the pie dough that's coming later down the line. Just rub. How long? For 37.2 seconds and not a moment longer. Seriously, you want to see about half the butter disappear into the flour and the rest hang around in little bits and bumps that are about the size of peas.

✦ Stick the butter and flour mixture in the freezer.

✦ Combine the Liquids and beat well.

✦ Remove the butter and flour mixture from its holding facility and dump the wet onto dry.

✦ Mix with a rubber or—preferably—silicone spatula. It'll bring things together fast and easy and as you'll soon see you're going to need it anyway.

> Notes:

> I've tried to figure out a better way to quantify this, but without a lab centrifuge and some sort of spectrometer thingy, I've yet to savvy a way.

> Never dump the dry into the wet. I've read instructions to do just that in many a cookbook and it just doesn't make sense. If the dry is on top, it will just sit there because it's light and aerated. If the wet goes on the dry, gravity will do part of the work—it's common sense.

✦ Knead the dough. The dough will be so wet when it's mixed that it will be almost a batter. So lay out a piece of waxed or parchment paper 18 to 20 inches long and dust it lightly with flour. Dump the dough in the middle and use the paper to shield your hands as you *fold* (not knead) the dough into a ½-inch to 1-inch thick rectangle.

Then use the ends of the wax paper to fold the dough as if you were making the world's biggest, stickiest tri-fold wallet.

Repeat this three times, being as gentle as possible.

The reason I don't want you to use your hands is that if you do, you'll almost certainly have to add flour to combat the resulting serious case of club hand.

Club hand

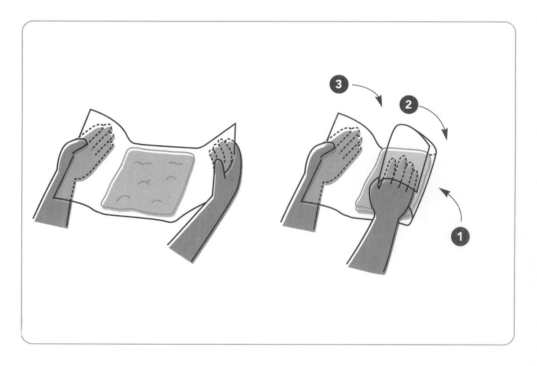

✦ Cut the biscuits. This is a free country and you can cut your biscuits to any size and shape you desire. I've seen square biscuits, I've seen tiny triangular biscuits…heck I've seen biscuits as big as a cat's head, but a 2-inch round biscuit beats them all. I'm convinced this size and shape makes for a better rise in the oven, which I imagine is because the surface-to-mass ratio is just right for the amount of lift generated by the leavening. Reshape the leftover dough, kneading as little as possible, and continue cutting out biscuits.

✦ As you cut out the biscuits, set them on an ungreased baking sheet, side by side, like this: Once they're all on the pan, lightly push down on the centers—just enough to barely dimple them (I use my middle finger to assure there's less pressure). Why? Physics. Since heat moves in from the outside, the middle of the biscuit usually sets last, so the center gets extra time to rise before setting, which is why so many biscuits look more like popovers than biscuits.

✦ Bake the biscuits in the preheated 450°F oven until golden brown and well...done, but not too done. This should take about 15 to 17 minutes. Why 450°? Because you want enough heat to generate steam and expand the gases inside the biscuit so that you get a good rise, but you don't want the outside of the biscuits to set before the rise is complete. I've found that 450° supplies a good balance of forces. And when they've got around 5 minutes left in the oven, brush the tops with melted butter.

✦ Turn the biscuits out into a basket lined with a clean kitchen towel and serve immediately with butter, jam...you know the drill. Without question, this is my personal best biscuit. If someone held a biscuit contest and I were the kind of guy who'd compete in a cooking contest, I'd enter these...and pity da po' fool that went up against me.

✦ I don't really know who has leftover biscuits, but if you do, they'll keep, but not for very long. Wrapped in foil, they'll start to go stale the next day—which is okay if you're going to split them and toast them as I strongly suggest you do, following the instructions in the note below. If you try to keep them in air-tight packaging, the biscuits will stay moist, but they'll also develop mold quickly—I'd say, after two days, max.

Note: Like whole wheat? Then replace half the all-purpose flour with whole-wheat flour. Don't worry, nothing bad will happen, although the biscuits won't keep quite as long.

The day after you make the biscuits, when they're just starting to stale a bit, split the leftovers in half with a fork—not a knife—and set them on a piece of foil folded up on the edges like a little tray. Top with butter and broil until brown and crispy. Now that's good...well, you know.

Yield: Eighteen 2-inch biscuits

TWICE COOKED

✦ ✦ ✦

In French, biscuit (bis-kwee) means "twice cooked" and refers to a wafer-like device that was designed to be consumed at sea (it had to be cooked twice to make it last). Although the word survives, the French meaning has absolutely nothing to do with the biscuit we know and love today.

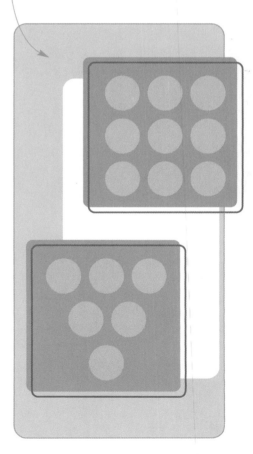

Hardware:

Digital scale

Dry measuring cups

Wet measuring cups

Measuring spoons

Food processor

Large mixing bowl

Box grater

Rubber or silicone spatula

Wax paper or parchment

Rolling pin

Bench scraper

2-inch round biscuit cutter

Baking sheet

Clean kitchen towel

Bread basket

Notes:

Pesto Dinner Biscuits

This recipe was developed so that I could have an excuse for eating biscuits at dinner…excuse me, supper. They go well with everything from simple roast chicken to standing rib roast. If you don't have lard, you may of course use shortening or more butter, but I think lard is better in this instance. A dinner biscuit should be more sophisticated and subtle. Lard provides a finer flake and a savory flavor that complements the pesto.

THE DRY GOODS

INGREDIENT	Weight		Volume	Count	Prep
All-purpose flour	270 g	9½ oz	2 cups		
Baking powder	14 g	½ oz	4 teaspoons		
Baking soda	2 g	<⅛ oz	¼ teaspoon		
Salt	4 g	⅛ oz	¾ teaspoon		

THE FAT

INGREDIENT	Weight		Volume	Count	Prep
Unsalted butter	42 g	1½ oz	3 tablespoons		frozen
Lard	14 g	½ oz	1 tablespoon		chilled

THE LIQUID

INGREDIENT	Weight		Volume	Count	Prep
Buttermilk	227 g	8 oz	1 cup		chilled

THE EXTRAS

INGREDIENT	Weight		Volume	Count	Prep
Pesto	128 g	4½ oz	½ cup		prepared or homemade
Flour for rolling out the dough					

✦ ✦ ✦ ✦ ✦

✦ Place an oven rack in position C and preheat the oven to 450°F.

✦ Assemble the dough via the **BISCUIT METHOD**. Combine the buttermilk and pesto before adding to the flour/fat mixture.

✦ Stir together until the dough pulls away from the sides of the bowl.

✦ Turn the dough out onto a sheet of lightly floured wax paper or parchment and roll out according to the directions on page 142.

✦ Using a 2-inch pastry cutter, cut out the biscuits and place them on an ungreased baking sheet, just touching each other. Reshape the leftover dough, kneading as little as possible, and continue cutting out biscuits and adding them to the baking sheet.

✦ Bake the biscuits for 15 to 17 minutes, or until golden brown.

✦ Remove the biscuits from the oven, place into a kitchen towel–lined basket, and serve immediately.

Yield: Eighteen 2-inch biscuits

Notes:

I like a 2-inch cutter, but a 3-inch works well, too.

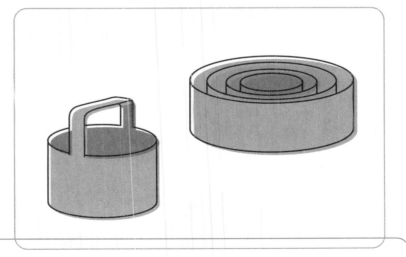

The sharper the sides, the better. You don't want to do this with an old can or a drinking glass. If you don't want to buy a biscuit cutter, just get a tin of pastry rings. That way, you can pick your size.

Hardware:

Digital scale

Dry measuring cups

Wet measuring cups

Measuring spoons

Chef's knife

Cutting board

Food processor

Large mixing bowl

Box grater

Rubber or silicone spatula

Wax paper or parchment

Rolling pin

Serrated bread knife

Baking sheet

Cooling rack

Notes:

Dried Cherry Scones

Scones are a tough concept for Americans to swallow (as are many things British). Just know that a scone is to a biscuit what a muffin is to a cake—drier and coarser in texture. Ultimately, scones are made for dunking.

THE DRY GOODS

INGREDIENT	Weight		Volume	Count	Prep
All-purpose flour	270 g	9½ oz	2 cups		
Baking powder	7 g	¼ oz	2 teaspoons		
Salt	4 g	⅛ oz	¾ teaspoon		
Sugar	64 g	2¼ oz	⅓ cup		

THE FAT

INGREDIENT	Weight		Volume	Count	Prep
Unsalted butter	85 g	3 oz	6 tablespoons		Frozen

THE LIQUID

INGREDIENT	Weight		Volume	Count	Prep
Heavy cream	177 g	6¼ oz	¾ cup		chilled
Eggs	100 g	3½ oz		2 large	beaten

THE EXTRAS

INGREDIENT	Weight		Volume	Count	Prep
Dried cherries	85 g	3 oz	⅓ cup		coarsely chopped
Flour for rolling out the dough					

✦ ✦ ✦ ✦ ✦

✦ Place an oven rack in position C and preheat the oven to 375°F.

✦ Assemble the dough via the **BISCUIT METHOD.** Add the dried cherries to the assembled dough and stir until just combined.

✦ Turn the mixture out onto lightly floured wax paper or parchment. Roll out into a 1-inch-thick round or rectangle, according to the directions on page 142.

✦ Using a serrated bread knife, cut the round or rectangle into 8 equal wedges. Place the wedges on an ungreased baking sheet about 1 inch apart.

✦ Bake the scones for 23 to 25 minutes, or until golden brown. Remove from the oven and place on a rack to cool. Serve at room temperature.

✦ Keep in an air-tight container for up to 3 days.

Yield: 8 scones

Notes:

since they're supposed to be tougher than biscuits, a little extra kneading won't hurt.

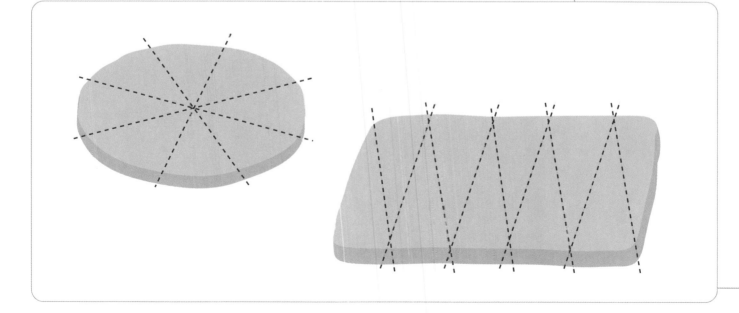

Blueberry Shortcake

This is a bit of a twist on the old standby, though in the spring I do use strawberries. I've been known to eat this to the point of illness…and I'm okay with that.

Hardware:

Digital scale

Dry measuring cups

Wet measuring cups

Measuring spoons

Medium mixing bowl

Potato masher

Wooden spoon

Food processor

Large mixing bowl

Box grater

Rubber or silicone spatula

Bench scraper

Baking sheet

Cooling rack

Serving spoon

Notes:

Because, obviously, this works with any kind of fruit that's in season.

THE FRUIT

INGREDIENT	Weight		Volume	Count	Prep
Fresh blueberries	539 g	1 lb 3 oz	4 cups		
Sugar	198 g	7 oz	1 cup		
Fresh lemon juice	57 g	2 oz	¼ cup		

THE DRY GOODS

INGREDIENT	Weight		Volume	Count	Prep
All-purpose flour	270 g	9½ oz	2 cups		
Baking powder	14 g	½ oz	4 teaspoons		
Salt	4 g	<⅛ oz	¾ teaspoon		
Sugar	64 g	2¼ oz	⅓ cup		

THE FAT

INGREDIENT	Weight		Volume	Count	Prep
Unsalted butter	85 g	3 oz	6 tablespoons		Frozen

THE LIQUID

INGREDIENT	Weight		Volume	Count	Prep
Heavy cream	177 g	6¼ oz	¾ cup		chilled
Eggs	50 g	1¾ oz		1 large	beaten

THE EXTRAS

Flour for rolling out the dough ✦ Whipped cream for serving

✦ ✦ ✦ ✦ ✦

✦ Make the blueberry mixture: dump half of the blueberries in a medium bowl with the cup of sugar and mash with a potato masher.

✦ Add the remaining berries and the lemon juice and stir. Let the mixture sit and steep at room temp (say between 40°F and 90°F). Give it an occasional stir while you're preparing the dough.

✦ Place an oven rack in position C and preheat the oven to 400°F.

✦ Assemble the dough via the **BISCUIT METHOD** and form it into a ball. Using a bench scraper, cut the ball in half, then fourths, and so on, until you have 8 equal parts. With your hands, form each piece of dough into a ball, pressing to flatten and form 2½-inch rounds that are ½ inch thick. Place each disk on an ungreased baking sheet about 1 inch apart.

✦ Bake for 20 minutes, or until golden.

✦ Remove the shortcakes from the oven and place on a rack to cool.

✦ When the shortcakes have cooled to room temperature, split them in half and place the bottom halves on dessert plates. Spoon some of the berries and juice over each. Replace the tops, spoon over some more berries, and finish with whipped cream.

✦ Shortcake is best when served immediately, but these will keep in an air-tight container for 3 days. Just be sure to cool to room temperature before wrapping them up.

Yield: 8 shortcakes

Notes:

Or you could weigh the whole thing and do the math . . . the choice is yours.

I cup of whipping cream whipped with 2 teaspoons of fine (not confectioners') sugar. Fine sugar (known as castor sugar in the UK) is easy to make—just spin granulated sugar in a blender or food processor for a minute.

Shortcake cut in half

Ice cream center

Another good use of shortcake.

Hardware:

Dutch oven

Digital scale

Dry measuring cups

Wet measuring cups

Measuring spoons

Wooden spoon

Food processor

Large mixing bowl

Box grater

Rubber or silicone spatula

Tablespoon

Pastry brush

Notes:

Frozen berries can be used as well, just increase the cooking time by 5 or 10 minutes.

Blackberry Grunt

This is a traditional New England dessert—think of it as a stove-top cobbler with very concentrated flavors. I'm told that a grunt is named for the sound it makes when the fruit bubbles up through the crust while it's baking. The original classic was made completely on the stove top—like chicken and dumplings, but with fruit. I finish mine in the oven so that the crust gets a dose of golden brown and delicious.

THE BLACKBERRY FILLING

INGREDIENT	Weight		Volume	Count	Prep
Water	227 g	8 oz	1 cup		
Sugar	210 g	7½ oz	1 cup		
Fresh blackberries	539 g	1 lb 3 oz	4 cups		
Ground ginger	3 g	⅛ oz	½ teaspoon		

THE DRY GOODS

INGREDIENT	Weight		Volume	Count	Prep
All-purpose flour	270 g	9½ oz	2 cups		
Baking powder	7 g	¼ oz	2 teaspoons		
Baking soda	2 g	<⅛ oz	¼ teaspoon		
Kosher salt	6 g	<¼ oz	1 teaspoon		

THE FAT

INGREDIENT	Weight		Volume	Count	Prep
Unsalted butter	57 g	2 oz	4 tablespoons	½ stick	Frozen

THE LIQUID

INGREDIENT	Weight		Volume	Count	Prep
Buttermilk	227 g	8 oz	1 cup		chilled

THE OPTIONS

Unsalted butter for brushing the top

Notes:

✦   ✦   ✦   ✦   ✦

✦ Make the blackberry filling. Place the water, sugar, blackberries, and ginger in a Dutch oven and place over medium heat. Bring to a simmer and cook the berry mixture, stirring occasionally, for 15 to 20 minutes, until the liquid is thick enough to coat the back of a spoon.

✦ Place an oven rack in position C and preheat the oven to 400°F.

✦ While you're cooking the filling, assemble the dough via the **BISCUIT METHOD** through step 5. You can stop after mixing; there's no rolling necessary.

✦ Drop the dough over the fruit mixture by the tablespoonful, evenly distributing it over the top. Bake for 15 to 20 minutes, just to brown the top of the dough. For a browner crust, brush with butter and broil until golden.

✦ Remove the pot from the oven and allow the grunt to rest 10 minutes before serving.

Yield: 6 to 8 servings

A Dutch oven is a large, heavy-gauge pot usually made out of cast iron that has a heavy, tight-fitting lid.

This is an optional step, but I personally like my crust golden brown.

Chicken and Dumplings

The dumplings in this recipe are drop-style like my mom used to make rather than the rolled, or flat, dumplings my mother-in-law makes (turn the page). It's not that I don't like my mother-in-law's dumplings; it's just that as far as I'm concerned, they're noodles. Not these…these are dumplings.

Hardware:

Digital scale
Dry measuring cups
Wet measuring cups
Measuring spoons
Chef's knife
Cutting board
Large soup pot
Food processor
Large mixing bowl
Box grater
Rubber or silicone spatula
1-ounce disher

Notes:

Either homemade stock or low-sodium canned stock will do. Of course, the homemade stuff will do better.

Poached is fine.

THE VEGETABLES AND CHICKEN

INGREDIENT	Weight	Volume	Count	Prep
Vegetable oil		2 tablespoons		
Onions		1 cup		finely diced
Celery		1 cup		finely diced
Carrots		½ cup		finely diced
Garlic		1 tablespoon		minced
Chicken stock		6 cups		
Cooked chicken		2 cups		cubed

THE DRY GOODS

INGREDIENT	Weight		Volume	Count	Prep
All-purpose flour	135 g	4¾ oz	1 cup		
Baking powder	7 g	¼ oz	2 teaspoons		
Baking soda	1 g	<⅛ oz	⅛ teaspoon		
Salt	3 g	<⅛ oz	½ teaspoon		
Freshly ground black pepper			to taste		
Dried sage	3 g	⅛ oz	1 teaspoon		
Dry mustard	3 g	⅛ oz	1 teaspoon		

THE FAT

INGREDIENT	Weight		Volume	Count	Prep
Unsalted butter	21 g	¾ oz	1½ tablespoons		frozen
Lard	7 g	¼ oz	2 teaspoons		chilled

THE LIQUID

INGREDIENT	Weight		Volume	Count	Prep
Buttermilk	113 g	4 oz	½ cup		chilled

✦ ✦ ✦ ✦ ✦

✦ Heat the oil in a large soup pot over medium-low heat. Add the onions, celery, carrots, and garlic and cook until tender, 5 to 7 minutes.

✦ Add the stock and cover the pot; increase the heat to medium and bring to a gentle boil. Reduce to a simmer and continue to cook for 7 to 10 minutes.

✦ Meanwhile, assemble the dumplings via the **BISCUIT METHOD.**

✦ Using the disher, scoop out the dough and roll it in your hands to form about 12 smooth balls.

✦ Drop the dumplings into the vegetable mixture, cover, and cook for 10 minutes over medium-low heat.

✦ Add the chicken, stir gently (just enough to evenly distribute the chicken so it heats through), and cook an additional 5 minutes. Season with salt and pepper and serve immediately.

Note: You can substitute 2 tablespoons chopped fresh herbs such as parsley, tarragon, thyme, or chives as an alternative to the sage and mustard.

Yield: 4 servings

Notes:

Hardware:

Digital scale

Dry measuring cups

Wet measuring cups

Measuring spoons

Food processor

Large mixing bowl

Rubber or silicone spatula

Rolling pin

Wax paper or parchment

Four clean kitchen towels

Large soup pot

Chef's knife

Cutting board

Large spoon

Notes:

Grandmom's Dumplings

In the interest of keeping my mother-in-law from turning my kitchen into something resembling *Kill Bill*'s House of Blue Leaves, I offer said mother-in-law's dumpling recipe as an alternative to the one on the previous page.

THE DRY GOODS

INGREDIENT	Weight		Volume	Count	Prep
All-purpose flour	270 g	9½ oz	2 cups		
Salt	6 g	<¼ oz	1 teaspoon		

THE FAT

INGREDIENT	Weight		Volume	Count	Prep
Vegetable shortening	71 g	2½ oz	⅓ cup		

THE LIQUID

INGREDIENT	Weight		Volume	Count	Prep
Milk	113 g	4 oz	½ cup		

THE CHICKEN

INGREDIENT	Weight		Volume	Count	Prep
Whole chicken	1.6 kg	3½ lb		1	cleaned

THE EXTRAS

Enough cold water to cover the chicken in the pot ✦ Salt to taste

✦ ✦ ✦ ✦ ✦

✦ Assemble the dumplings according to the **BISCUIT METHOD,** adding only enough milk to make a rollable dough (it may take more milk, it may take less).

✦ Divide the dough into two equal balls. Roll out each ball of dough to a thickness of $\frac{1}{16}$ to $\frac{1}{8}$ inch on wax paper or parchment that has been dusted with flour.

✦ Lay a clean kitchen towel on the counter and then move the dough—on the waxed paper—on top of the towel. Top the dough with a second clean kitchen towel. Repeat with the second ball of dough, then leave the dough to dry for at least 8 hours (you can make the dough in the morning and leave it to dry till dinnertime).

✦ Meanwhile, cook the chicken. Place the whole chicken in a large soup pot, cover with water, and add salt to taste. Bring the water to a boil, then reduce to a simmer. Cook until the chicken is fully poached, about 40 minutes.

✦ Remove the chicken from the pot to cool; reserve the broth. When the chicken has cooled, remove the meat from the bone in chunks and return the meat to the broth.

✦ Bring the broth and chicken to a rolling boil. Cut the dough into $\frac{1}{2}$-inch strips, break off short pieces, and drop them into the boiling broth. *Do not stir the dumplings into the broth.* When all the dough has been used, gently push all the dumplings down into the broth with a spoon.

✦ When the dumplings have cooked through, about 10 minutes, ladle them into soup bowls with the chicken pieces and broth.

Yield: Serves 4 to 6

Notes:

I think thinner is better here, but you probably won't be able to manage $\frac{1}{16}$TH—so start at $\frac{1}{8}$TH inch and work your way down.

Hardware:

Digital scale

Dry measuring cups

Wet measuring cups

Measuring spoons

Food processor

2 large mixing bowls

Box grater

Rubber or silicone spatula

Rolling pin

Rolling pin bands

Pizza or pastry cutter

Metal or offset cake spatula

Parchment paper

Baking sheet

Pastry brush

Cooling rack

Notes:

A Superior Saltine

Yes, I realize that store-bought saltines are cheap and readily available, and that's why no one makes them. But if we did make them every now and then, we'd realize that in the rock-scissors-paper game of life, easy and tasty beats out cheap and readily available every time. These crackers are elegant enough to be served as the bread at a fancy-shmancy dinner party.

THE DRY GOODS

INGREDIENT	Weight		Volume	Count	Prep
All-purpose flour	270 g	9½ oz	2 cups		
Salt	6 g	<¼ oz	1 teaspoon		

THE FAT

INGREDIENT	Weight		Volume	Count	Prep
Unsalted butter	57 g	2 oz	4 tablespoons	½ stick	frozen

THE LIQUID

INGREDIENT	Weight		Volume	Count	Prep
Eggs	50 g	1¾ oz		1 large	beaten
Milk	113 g	4 oz	½ cup		

THE EXTRAS

INGREDIENT	Weight		Volume	Count	Prep
Unsalted butter for brushing on top	28 g	1 oz	2 tablespoons		melted

Flour for rolling out the dough ✦ Kosher salt for sprinkling on top

Salt fans, this is the perfect place to show off your best stuff. I like Welsh sea salt but kosher's good, too. Just don't use table salt . . . please.

✦ ✦ ✦ ✦ ✦

✦ Place an oven rack in position C and preheat the oven to 325°F.

✦ Assemble the dough via the **BISCUIT METHOD,** but let the fat sit for a few minutes after you grate it into the flour, before rubbing it in. Your crackers will be less crumbly, and they won't try to rise.

✦ Turn the dough out onto a lightly floured surface and roll to about $\frac{1}{16}$ inch.

✦ Cut into the shapes of your choice with a pizza/pastry cutter (I usually go with squares).

✦ Use a metal spatula or offset cake spatula to move the crackers to a baking sheet lined with parchment paper.

✦ Bake for 30 to 35 minutes, or until golden. Brush with the melted butter and sprinkle with the salt.

✦ Cool on a rack and store in an airtight container for up to 1 month.

Yield: 45 to 50 crackers

CLARIFIED BUTTER

✦ ✦ ✦

Since water will make the crackers mushy, use either high-quality butter (such as plus gras) or clarify your butter. To clarify: Melt a stick of butter in a small, heavy saucepan (I usually use a heavy metal 1 cup measuring cup) over low heat and slowly cook until the bubbling ceases and the liquid turns clear, 10 to 15 minutes. Strain and cool, being sure to leave any solids in the bottom of the pan. Or, once the butter has cleared, remove from heat and add 2 inches of hot tap water. Since butter is less dense than water, the now clarified butter will float to the top. A few hours in the fridge will solidify the butter. Use or wrap in wax paper and foil and refrigerate, or you can keep it frozen for up to 2 months.

✦ ✦ ✦ ✦ ✦

Variation

Here's a variation that's perfectly suited to an elegant dinner. Crank your oven to 350°F. Melt a stick of butter in a metal bowl (right on the stove top). Dip the cracker in the butter, lay it on a rack, and bake until toasty brown. Cool and serve on a bread plate just like fancy French bread. Yum.

One of my personal heroes, Shirley O. Corriher, and her husband, Arch (who plays my eighth-grade science teacher on TV), occasionally treat my wife and me to dinner at their club. That's where I first met homemade saltines baked in butter. I usually eat about a dozen.

Okay, $\frac{1}{16}$TH is kinda tricky. So take the tricky out by purchasing a set of rolling pin bands. They're rubber bands in different thicknesses that you place on either end of your rolling pin to assure that you roll thin but not too thin. Or you can roll the dough to $\frac{1}{4}$ inch, cut it into long strips, and run them through a pasta roller.

Hardware:

Digital scale

Dry measuring cups

Wet measuring cups

Measuring spoons

Food processor

2 large mixing bowls

Box grater

Rubber or silicone spatula

Rolling pin

Rolling pin bands

Pizza or pastry cutter

Metal or offset cake spatula

Baking sheet

Pastry brush

Cooling rack

Notes:

Whole Wheat Crackers

I won't say this is my favorite application in this book, but if I did, I wouldn't be lying. If you're a Wheat Thins fan, trust me…you want to make this cracker. I make a batch a week every Sunday when I'm home, and I'm usually out of them by Tuesday.

THE DRY GOODS

INGREDIENT	Weight		Volume	Count	Prep
Whole wheat flour	142 g	5 oz	1 cup		
All-purpose flour	142 g	4¾ oz	1 cup		
Sugar	50 g	1¾ oz	¼ cup		

THE FAT

INGREDIENT	Weight		Volume	Count	Prep
Unsalted butter	57 g	2 oz	4 tablespoons	½ stick	

THE LIQUID

INGREDIENT	Weight		Volume	Count	Prep
Milk	113 g	4 oz	½ cup		
Eggs	50 g	1¾ oz		1 large	

THE EXTRAS

INGREDIENT	Weight		Volume	Count	Prep
Unsalted butter for brushing on top	28 g	1 oz	2 tablespoons		melted

Flour for rolling out the dough ✦ Kosher salt for sprinkling on top ✦ Parchment for the baking sheet

Notes:

✦ ✦ ✦ ✦ ✦

✦ Place an oven rack in position C and preheat the oven to 325°F.

✦ Assemble the dough via the **BISCUIT METHOD,** allowing the fat to sit for a few minutes after grating it onto the flour.

✦ Roll out the dough following the method for A Superior Saltine (page 157).

✦ Use a metal spatula or offset cake spatula to move the crackers to a baking sheet lined with parchment paper.

✦ Bake for 30 to 35 minutes, or until golden.

✦ Brush with the melted butter and sprinkle with the salt.

✦ Cool on a rack, store in an airtight container for 1 month.

Yield: 45 to 50 crackers

My favorite thing to put on these is . . . actually, I like them plain and unencumbered.

✦ ✦ ✦ ✦ ✦

Variation

Before baking, try sprinkling with caraway seeds or lightly crushed cumin seeds. They can also handle a big dose of pepper.

Hardware:

Digital scale

Dry measuring cups

Measuring spoons

Chef's knife

Cutting board

Food processor

Large mixing bowl

Box grater

Notes:

Straightforward Streusel

Streusel is German for "sprinkle," and that's exactly what you do with this tasty stuff. If you pack it tightly, it turns hard and nasty, so sprinkle, even if you sprinkle it two inches deep…which I do from time to time.

Streusel is one of the most versatile quick combos in the kitchen because it works and plays well with just about any fruit. It can be baked and broken up into something very much like granola. In fact, you can make a darned tasty streusel out of granola as long as it isn't too crunchy.

THE DRY GOODS

INGREDIENT	Weight		Volume	Count	Prep
All-purpose flour	32 g	1⅛ oz	¼ cup		
Light brown sugar	42 g	1½ oz	¼ cup		
Salt	2 g	<⅛ oz	¼ teaspoon		
Ground cinnamon	2 g	<⅛ oz	¼ teaspoon		
Almonds	42 g	1½ oz	¼ cup		toasted and chopped

THE FAT

INGREDIENT	Weight		Volume	Count	Prep
Unsalted butter	42 g	1½ oz	3 tablespoons		Softened

✦ ✦ ✦ ✦ ✦

✦ Make the streusel via the **BISCUIT METHOD** through step 3 and you're done.

✦ The mixture should be loose and…well, I think it's sorta like potting soil, only not so dark and a lot more tasty. Use immediately or cover and refrigerate for up to a month. (Since there's a lot of fat in there, the streusel will pick up random flavors if the seal isn't tight.)

✦ Possible uses? Pack some streusel in the bottom of a buttered baking dish, top with slices of fruit, top with more streusel, and bake at 350°F until you can see juices bubbling up through the crust.

The Biscuit Method/ Pie Variation

Despite the fact that it's as simple as the day is long, pie crust remains the Cerberus of the baking world. Many beginners take one gander at it, turn tail, and run. That's because so many of the crucial points of action are described by words that just don't mean a heck of a lot. There are a lot of *"just enoughs"* and *"not too muches"* and other verbal voodoos that can ruin your day. Some of that goes with the territory, I suppose, but if you pay close attention, you shouldn't have to go at it more than three times before you nail it.

Notes:

◆ Place the other pan (inverted) on top of the dough and push gently downward. You should now have one pie dough sandwiched between two pie tins that are both facing down. Odds are good there will be at least an inch of pie dough sticking out of this union all the way around. Use your paring knife to trim the ragged edge away, leaving a half inch of dough all the way around, if possible.

◆ Carefully flip the whole thing back over again.

◆ Remove pan number 1 and—behold—a perfect pie dough perfectly positioned in a pie plate.

Note: This isn't nearly as hard to do as it is to read. That's why there are pictures.

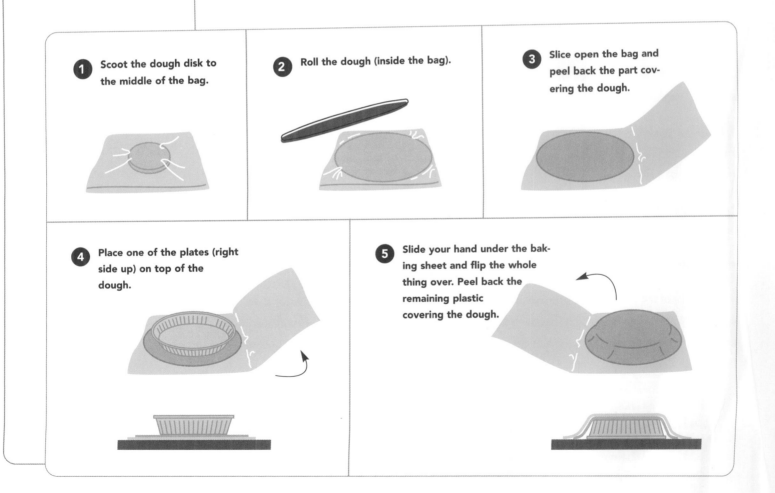

1 Scoot the dough disk to the middle of the bag.

2 Roll the dough (inside the bag).

3 Slice open the bag and peel back the part covering the dough.

4 Place one of the plates (right side up) on top of the dough.

5 Slide your hand under the baking sheet and flip the whole thing over. Peel back the remaining plastic covering the dough.

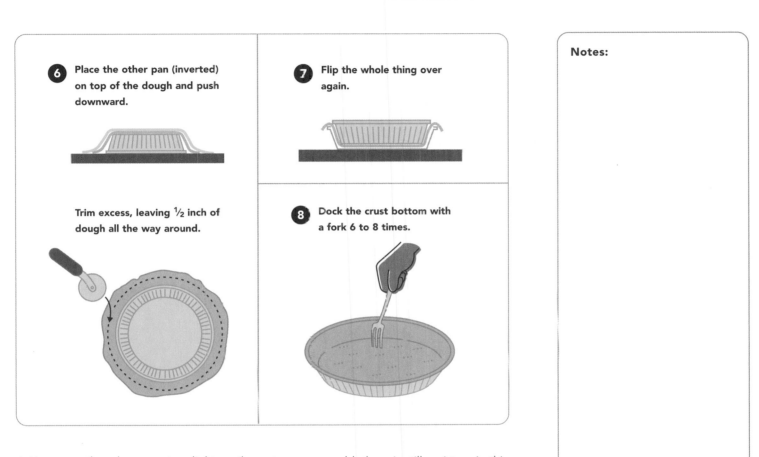

6 Place the other pan (inverted) on top of the dough and push downward.

7 Flip the whole thing over again.

Trim excess, leaving ½ inch of dough all the way around.

8 Dock the crust bottom with a fork 6 to 8 times.

Notes:

✦ Now even though we went as light on the water as we could, there is still moisture in this dough, and when it faces the heat of the oven it will turn to steam. That steam could push our dough skyward. The solution? Docking.

I have no idea why they call it docking because it's got nothing to do with any of the connotations I associate with that word. In pastry parlance to dock means "to poke full of holes." So, take a regular ol' fork and poke the crust bottom 6 to 8 times. That should do the job.

✦ Cover the pie with a 13 x 13-inch piece of parchment paper and fill with the dry beans…just to make sure the dough remains flat. Bake for 10 minutes, then remove the parchment and beans and continue baking until golden, 10 to 15 minutes. Remove from the oven and place on a rack to cool completely before filling.

Yield: One 9-inch pie crust

One of my earliest memories in life is when I "docked" my dad's Naugahyde recliner with a drill bit I'd found in the garage.

Hardware:

Digital scale

Dry measuring cups

Wet measuring cups

Measuring spoons

Peeler

Chef's knife

Cutting board

Box grater

Whisk

Large cast-iron skillet

Wooden spoon

Food processor

Spritz bottle

Sheet pan or cookie sheet

Rolling pin

Pastry brush

Cooling rack

Notes:

Free-Form Apple Pie

This is my favorite kind of pie to make because it requires no pie pan. Although this is technically a galette, or free-form tart, this here's America, and in America we eat apple pie, gosh darn it. Besides, I've never had any patience for lattice-top pies. This is far more satisfying, and the traditional cheddar cheese is built in. If the inclusion of vinegar seems odd, trust me when I say it brings more than potato chips to the party.

THE DRY GOODS

INGREDIENT	Weight		Volume	Count	Prep
All-purpose flour	337 g	12 oz	$2^{1}/_{2}$ cups		
Stone-ground cornmeal	64 g	$2^{1}/_{4}$ oz	$^{1}/_{2}$ cup		
Sugar	46 g	$1^{1}/_{2}$ oz	3 tablespoons		
Kosher salt	4 g	$< ^{1}/_{4}$ oz	1 teaspoon		

THE FAT

INGREDIENT	Weight		Volume	Count	Prep
Unsalted butter	227 g	8 oz	1 cup	2 sticks	diced

THE LIQUID

INGREDIENT	Weight		Volume	Count	Prep
Apple juice	57 g	2 oz	$^{1}/_{4}$ cup		
Ice water	57 g	2 oz	$^{1}/_{4}$ cup		

THE FILLING

INGREDIENT	Weight		Volume	Count	Prep
Pink Lady, Braeburn, or other tart, firm apples	340 g	12 oz		2	peeled, cored, and thinly sliced
Cider vinegar	43 g	1½ oz	3 tablespoons		
Sugar	50 g	1¾ oz	¼ cup		
Grated nutmeg			pinch		
Ground cinnamon	2 g	<⅛ oz	¼ teaspoon		
Unsalted butter	28 g	1 oz	2 tablespoons		
Cheddar cheese	57 g	2 oz	½ cup		shredded
Flour	3 g	<⅛ oz	1 teaspoon		

THE EXTRAS

INGREDIENT	Weight		Volume	Count	Prep
Pound cake	128 g	4½ oz	1½ cups		roughly chopped into ½-inch cubes
Unsalted butter	28 g	1 oz	2 tablespoons		cubed
Eggs	50 g	1¾ oz		1 large	beaten with 1 tablespoon water
Sugar	3 g	<⅛ oz	½ teaspoon		

◆ ◆ ◆ ◆ ◆

✦ Mix the apple juice and the ice water and pour into a spritzer bottle.

✦ Assemble the dough via the **BISCUIT METHOD/PIE VARIATION,** incorporating the butter in 2 doses and using only as much of the juice/water combination as necessary.

✦ Warm a cast-iron skillet over medium heat. Add the apples to the pan and toss for 2 minutes.

✦ Add the vinegar and stir for 30 seconds.

✦ Add the sugar and cook until the apples have softened, 2 to 3 minutes.

Notes:

Notes:

✦ Add the nutmeg, cinnamon, and butter and allow the butter to melt slowly.

✦ Fold in the cheddar and remove from the heat.

✦ Sprinkle on the flour and stir to combine well.

✦ Cool to room temperature.

✦ While the apple mixture is cooling, place an oven rack in position C and preheat the oven to 400°F.

✦ Roll the dough into a ¼-inch-thick disk. Don't worry about getting it perfectly round. Carefully roll it up onto the rolling pin and transfer it to the baking sheet.

✦ Place the pound cake pieces in the middle of the dough, leaving a 3-inch margin of crust on all sides.

✦ Spoon the apple mixture over the pound cake and top the apples with the cubed butter. Lift the excess crust onto the filling and repeat in a clockwise fashion until a top lip has formed around the edge of the whole pie. Brush the pie with the egg wash and sprinkle the crust with the sugar.

✦ Bake for 30 to 35 minutes, or until the filling begins to bubble and the crust is golden brown.

✦ Remove the pie from the sheet pan immediately and place on a rack to cool. Serve at room temperature.

Yield: One 12-inch free-form pie

Savory Pie Crust

This pie crust is—as its name says—savory, and very flavorful. Blind-baked, it's the perfect construction platform for a quiche or any other savory pie, or you could use it to make the Apple, Fennel, and Onion Pie on pages 172–173.

THE DRY GOODS

INGREDIENT	Weight		Volume	Count	Prep
All-purpose flour	270 g	9½ oz	2 cups		
Salt	7 g	¼ oz	1½ teaspoons		
Freshly ground black pepper	5 g	<¼ oz	1 teaspoon		

THE FAT

INGREDIENT	Weight		Volume	Count	Prep
Unsalted butter	113 g	4 oz	8 tablespoons	1 stick	

THE LIQUID

INGREDIENT	Weight		Volume	Count	Prep
Ice water	57 g	2 oz	¼ cup		

THE EXTRAS

INGREDIENT	Weight		Volume	Count	Prep
Parmesan cheese	50 g	1¾ oz	½ cup		grated

◆　◆　◆　◆　◆

◆ Assemble the dough via the **BISCUIT METHOD/PIE VARIATION,** adding the Parmesan to the Dry Goods just before you add the Liquid.

◆ Roll and blind-bake the dough for a traditional pie crust following the step-by-step instructions on pages 164–165.

Yield: Two 9-inch pie crusts

Hardware:

Digital scale

Dry measuring cups

Wet measuring cups

Measuring spoons

Food processor

Spritz bottle

Large zip-top bag

Sheet pan or cookie sheet

2 matching 9-inch pie tins

Rolling pin

Paring knife

Kitchen fork

Parchment paper

Approximately 32 ounces dried beans

Cooling rack

Notes:

Hardware:

8 x 8-inch glass baking dish

Digital scale

Dry measuring cups

Measuring spoons

Chef's knife

Cutting board

Food processor

Spritz bottle

Medium mixing bowl

Serving spoon

Notes:

Using the frozen fruit right out of the freezer is the secret to keeping their flavorful juice—and the extra dough on the top and bottom soaks up the juices.

Peach and Rhubarb Cobbler

Peaches are my favorite fruit, hands down...and I love cooking with rhubarb because it's poison. Well, the leaves contain oxalates, a kind of poison. But don't worry, the rhubarb we eat comes from the stem and has been completely cleared for human consumption. Anyway, this is my favorite cobbler of all time. Don't eat it all at one sitting because it gets better with age.

THE DRY GOODS

INGREDIENT	Weight		Volume	Count	Prep
All-purpose flour	270 g	9½ oz	2 cups		
Salt	6 g	<¼ oz	1 teaspoon		
Sugar	28 g	1 oz	2 tablespoons		
Lime zest	4 g	<⅛ oz	1 tablespoon	1 lime	

THE FAT

INGREDIENT	Weight		Volume	Count	Prep
Unsalted butter	128 g	4½ oz	9 tablespoons		chilled and cut into small pieces
Lard	42 g	1½ oz	3 tablespoons		chilled

THE LIQUID

INGREDIENT	Weight		Volume	Count	Prep
Ice water	43 g	1½ oz	3 tablespoons		

THE FRUIT

INGREDIENT	Weight		Volume	Count	Prep
Sugar	210 g	7½ oz	1 cup		
Cornstarch	14 g	½ oz	2 tablespoons		
Salt	2 g	<⅛ oz	¼ teaspoon		
Frozen peaches	16-ounce bag				unthawed
Frozen rhubarb	16-ounce bag				unthawed
Fresh lime juice	11 g	⅜ oz	1 tablespoon	1 lime	

THE EXTRAS

Unsalted butter for the dish

Notes:

✦ ✦ ✦ ✦ ✦

✦ Place an oven rack in position C and preheat the oven to 375°F. Butter the baking dish.

✦ Assemble the dough via the **BISCUIT METHOD/PIE VARIATION** and place it in the refrigerator while you put the fruit together.

✦ To make the fruit, in a medium bowl whisk together the sugar, cornstarch, and salt, then add the peaches, rhubarb, and lime juice.

✦ Remove the dough from the refrigerator, cut it in half, then pinch off walnut-sized pieces from half of the dough and drop the pieces onto the bottom of the prepared dish.

✦ Top the dough that's on the baking dish with the fruit mixture, then drop the remaining half of the dough randomly, in big spoonfuls, on top of the fruit.

✦ Bake, uncovered, for 1¼ hours, until the dough is cooked through and starting to turn golden, then place under the broiler for 3 minutes just to brown the top.

✦ Remove from the broiler and let stand for 15 minutes before serving.

Yield: Serves 6 to 8, unless I'm one of the diners—then it serves 3

You can use a toothpick to check for doneness here, just be careful not to poke through to the fruit. And if you can lift it up from below with a spatula and the dough is solid, you're good to go.

Apple, Fennel, and Onion Pie

A savory twist on the free-form pie, very nice alongside a salad or a bowl of soup.

THE DOUGH

INGREDIENT	Weight		Volume	Count	Prep
Savory Pie Crust			1 recipe		see page 169

THE FILLING

INGREDIENT	Weight		Volume	Count	Prep
Grainy mustard			1 tablespoon		
Baking apple (Gala, Cortland, or Braeburn)	163 g	5¾ oz	1½ cups	1	cored and sliced into ¼-inch thick wedges
Small fennel bulb	170 g	6 oz	2 cups	1	cored and sliced into ¼-inch thick wedges
Small onion	170 g	6 oz	1½ cups	1	cut into ¼-inch thick wedges
Fresh thyme			1 teaspoon		picked

THE EXTRAS

INGREDIENT	Weight		Volume	Count	Prep
Unsalted butter	28 g	1 oz	2 tablespoons		melted

Hardware:

Digital scale

Dry measuring cups

Wet measuring cups

Measuring spoons

Chef's knife

Cutting board

Food processor

Spritz bottle

Rolling pin

Sheet pan or cookie sheet

Parchment paper

Pastry brush

Cooling rack

Notes:

Notes:

✦ ✦ ✦ ✦ ✦

✦ Place an oven rack in position C and preheat the oven to 375°F.

✦ Assemble the ingredients of the Savory Pie Crust recipe (see page 169) via the **BISCUIT METHOD/PIE VARIATION.**

✦ To make the free-form crust, roll the dough into a ¼-inch-thick disk. Don't worry about getting it perfectly round. Carefully roll it up onto the rolling pin and transfer it to the baking sheet.

✦ Spread the mustard in the center of the dough.

✦ Randomly but evenly disperse the wedges of apple, fennel, and onion.

✦ Sprinkle the top with the thyme.

✦ Fold the excess crust onto the filling and repeat in a clockwise fashion until a top lip has formed around the edge of the whole tart (see the illustration on page 168).

✦ Brush the tart with the melted butter and bake for 50 to 55 minutes, or until the crust is golden and the filling is soft.

Yield: Serves 6 to 8

The Creaming Method

If you've ever made cookies or cakes from a recipe, odds are good you've come across this sentence: "Cream butter (or shortening) with sugar until light and fluffy." The reason is this: Creaming creates products with a fine, regular texture and a soft, moist (but not necessarily tender) tooth. The problem is that most recipes fail to explain *why* you need to "cream" anything, nor do they tell you how to manage the creaming process.

directly to fat. Since fats and water don't get along very well, batters mixed this way tend to come together slowly. To get around this, mix the eggs together first so that the water in the egg whites can hook up with the emulsifiers in the yolks. Add them in a steady stream and the eggs will be absorbed in no time.

Drop the mixer speed to low and add the dry goods slowly, alternating the addition of the dry goods with any other liquid ingredients. Then stir in any bits and pieces like chocolate chips at the end.

Depending on the amount of liquid involved, the mixture produced may be a pourable batter (a cake) or a thick paste (a chocolate chip cookie). So your final step would be to pour the batter into or "drop" the cookies onto a pan and bake.

It's the Temperature of the Fat

THE SUCCESS OF CREAMING depends in no small part on the temperature of the fat. Most recipes use the term "room temperature" as though it actually meant something. Depending on the season, my kitchen can range from 65° to 80°F, and that's if I break down and use the air conditioner.

So, if butter is the fat in question, I break it out of the fridge and counter it until it hits between 65° and 70°F, maintaining the temp with a probe thermometer.

If shortening is the fat, I like 70°F or lower. During warm summer months, I keep my shortening tightly lidded in the refrigerator so that it's always ready. The nice thing about shortening is that it's almost as easy to work with at 50° as 70°F. If you just can't bring yourself to refrigerate shortening, you can always stick it in ice water.

Shortening can be measured by volume and chilled at the same time by being placed in a measured amount of ice water. Simply submerge shortening with a skewer or fork and measure by displacement. Leave in water 5 minutes. Fish it out, dry on a paper towel, and go to work.

Tightly lidded because fats are quite good at picking up random flavors.

2 cups H$_2$O

Toothpick

Shortening

Proper Pan Prep

FEW THINGS CAN RUIN YOUR DAY LIKE a cake or loaf or muffin that refuses to let go of its cooking vessel. Proper pan prep can prevent such stubbornness. I grease and line every pan I bake with, even if the application doesn't call for it. It only takes a couple of minutes to take out an insurance policy against such disasters…isn't your baked good worth it?

I use one of the following three methods, depending on the target food.

Baker's Joy. This is a spray lube that actually contains flour. So when some recipe (none herein, I hope) demands that you grease and flour, you can do it all with the push of a button. It's the only manufactured pan release application I use in baking—and it's especially good for Bundt pans, muffin tins, and any pan with hard-to-get spots. If you can't get your hands on some, then there's…

AB's Kustom Kitchen Lube. Yep, make this stuff up right there at home. Just toss 2 cups of shortening into the ol' stand mixer with 1½ cups of all-purpose flour and mix it on low, just until the shortening sucks up the flour. Then hike the speed up to medium to aerate it a bit for easy application. Store it in a resealable plastic container and use it to lube up anything and everything. It's good on just about any type of pan.

Lube pan with shortening. Why shortening and not butter? Because butter contains two things I don't want under or around the sides of my baked good: water and protein. Butter may only contain some 15 percent water, but as that water changes to steam, it increases in volume by about a thousand times. That steam's got to go someplace and it can wreak havoc as it does. Butter also contains proteins…so does Elmer's glue. Enough said.

Parchment. I paper just about everything. Here's how:

No, wax paper will not do. Parchment is impregnated with silicone, which is a wonderfully nonreactive substance (that means it doesn't stick), and it's very stable. Wax paper is nonreactive only as long as the wax isn't melted. Line your cake and bread pans with wax paper only if you're looking for a nice crayon flavor to infuse your foods.

To line a round cake pan:

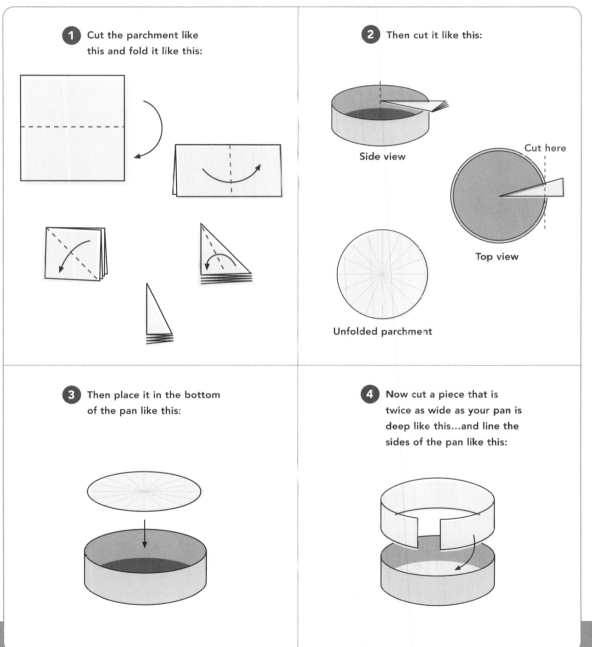

1 Cut the parchment like this and fold it like this:

2 Then cut it like this:

Side view

Cut here

Top view

Unfolded parchment

3 Then place it in the bottom of the pan like this:

4 Now cut a piece that is twice as wide as your pan is deep like this...and line the sides of the pan like this:

In the case of squares or rectangles, I generally make a sling, leaving two of the sides (the short ones on a loaf pan) greased only.

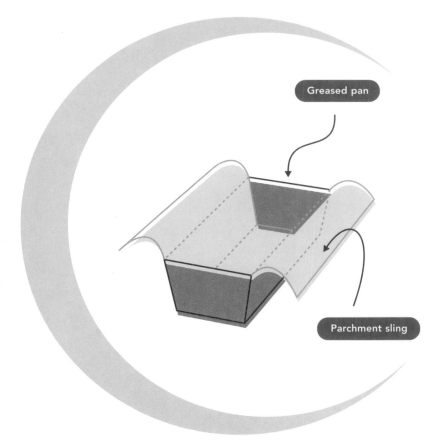

Greased pan

Parchment sling

Another reason to love parchment paper: Often the starch and protein inside cookies hasn't set all the way when it's time to evacuate the oven. In these cases attempting removal via spatula can end in misshapen cookies or worse. By baking on parchment, you can simply slide the entire batch off to a cooling rack without touching a single cookie. That way the pan is free for another load and the cookies are safe and sound.

Parchment

Cookies cool

The two exceptions. Soufflé dishes get buttered then dusted with flour or grated cheese (see page 274) because the foam needs something to cling to as it climbs. Pans used for angel food cakes aren't lubed at all, because the fat would break down the foam structure. That's why you generally have to cut angel food cakes out of the pan with a knife...hence the two-piece tube pan.

Hardware:

Digital scale

Dry measuring cups

Wet measuring cups

Measuring spoons

Nutmeg grater or microplane

Chef's knife

Cutting board

12-cup Bundt pan

Food processor

Large mixing bowl

Whisk

Stand mixer with paddle attachment

Rubber or silicone spatula

Cooling rack

Notes:

This is the classic Bundt. The fancier flower and castle shapes usually have a 10-cup capacity.

I like Galas.

Apple Cake

Another really old recipe from the Brown family books.

THE CREAMED

INGREDIENT	Weight		Volume	Count	Prep
Unsalted butter	227 g	8 oz	1 cup	2 sticks	softened but not melting (70°F)
Sugar	397 g	14 oz	2 cups		

THE EGGS

INGREDIENT	Weight		Volume	Count	Prep
Eggs	100 g	3½ oz		2 large	beaten
Vanilla extract	14 g	½ oz	1 tablespoon		

THE DRY GOODS

INGREDIENT	Weight		Volume	Count	Prep
All-purpose flour	270 g	9½ oz	2 cups		
Baking soda	12 g	½ oz	2 teaspoons		
Salt	6 g	<¼ oz	1 teaspoon		
Ground cinnamon	3 g	<⅛ oz	1 teaspoon		
Freshly grated nutmeg	3 g	<⅛ oz	1 teaspoon		

THE EXTRAS

INGREDIENT	Weight		Volume	Count	Prep
Apples	454 g	1 lb	4 cups		Cored and roughly chopped into 1-inch pieces or smaller; don't peel
Pecans			2 cups		chopped
Baker's Joy or AB's Kustom Kitchen Lube for the pan					

✦ ✦ ✦ ✦ ✦

✦ Place an oven rack in position C and preheat the oven to 275°F.

✦ Prepare the Bundt pan (see pages 180–183) and set aside.

✦ Assemble the ingredients via the **CREAMING METHOD,** folding in the apples and pecans last.

✦ Pour the batter into the Bundt pan and bake until the internal temperature reaches 212°F, about 1 hour and 45 minutes.

✦ Cool in pan for 15 minutes, then remove to rack to cool completely.

✦ Wrapped tightly, this will keep for 5 days.

Yield: Serves 12

Hardware:

9 x 3-inch round cake pan or
9 x 13-inch rectangular cake pan

Food processor

Digital scale

Dry measuring cups

Wet measuring cups

Measuring spoons

Large mixing bowl

Whisk

Stand mixer with paddle attachment

Rubber or silicone spatula

Cooling rack

Chef's knife

Cutting board

Notes:

Unlike most creaming procedures in which the sugar disappears into the butter, in this case the butter will completely and uniformly disappear into the sugar. That's okay.

If you want a softer cake, you can substitute 10 ounces of cake flour. I like the A-P version better—it's meatier.

Fudge Cake

You'll be sure that you've done something wrong when you've made this batter—it's very loose for a cake. Don't worry. It will be fine. In fact, this cake is so good that it doesn't even need frosting, but I've given you a recipe anyway. And if you liked Hostess Cupcakes when you were growing up, this recipe comes pretty close. Just substitute a muffin tin for the cake pan, fill the cups (with or without paper inserts) three-quarters full, and bake for 15 to 20 minutes in a 350°F oven, rotating the pan half-way through. You'll get about 2 dozen cupcakes.

THE CREAMED

INGREDIENT	Weight		Volume	Count	Prep
Unsalted butter	113 g	4 oz	8 tablespoons	1 stick	
Brown sugar	383 g	13½ oz	2¼ cups		

THE EGGS

INGREDIENT	Weight		Volume	Count	Prep
Eggs	150 g	5¼ oz		3 large	
Vanilla extract	7 g	¼ oz	1½ teaspoons		

THE DRY GOODS

INGREDIENT	Weight		Volume	Count	Prep
Unsweetened baking chocolate	85 g	3 oz		3 squares	
All-purpose flour	305 g	10¾ oz	2¼ cups		
Baking soda	12 g	<½ oz	2 teaspoons		
Salt	6 g	<¼ oz	1 teaspoon		

THE LIQUID

INGREDIENT	Weight		Volume	Count	Prep
Sour cream	227 g	8 oz	1 cup		full fat
Water	227 g	8 oz	1 cup		boiling

THE EXTRAS

Shortening and parchment paper for the pan

✦ ✦ ✦ ✦ ✦

✦ Place an oven rack in position C and preheat the oven to 350°F.

✦ Prep a 9 x 3-inch-deep round cake pan or a 9 x 13-inch rectangular pan (see pages 180–183) and set aside.

✦ Pulverize the chocolate in a food processor until there is nothing but chocolate dust and small chunks. Add the remaining Dry Goods and pulse several times until the chocolate and flour are completely homogenized. Set aside.

✦ Beat together the eggs and the vanilla extract. Set aside.

✦ Assemble all the ingredients, except the water, via the **CREAMING METHOD,** alternating three doses of the Dry Goods with two doses of the sour cream.

✦ With the stand mixer on low, add the boiling water and continue mixing until a loose but uniform batter (with no lumps) is achieved.

✦ Pour the batter into the prepared pan and bake for ½ hour at 350°F, then reduce the oven to 300°F and bake for another ½ hour. The cake is done when the internal temperature has reached 175° or 180°F. Although a toothpick inserted at mid-radius (halfway between the edge and the center) will come out clean, that same toothpick plunged into the center of the cake will come out gooey. Don't mind that…the cake will continue to cook as it cools.

✦ Remove the cake from the oven and let it sit in the pan for 15 minutes before turning it out onto a cooling rack. Allow the cake to cool completely before frosting.

✦ If you've chosen the round option, split the cake into two or three layers (see page 131) before frosting. If you have chosen the rectangular option, frost and then cut into squares.

Yield: Serves 8 to 12

Notes:

Don't forget the frosting.

Hardware:

Digital scale

Wet measuring cups

Dry measuring cups

Measuring spoons

Food processor

8-inch square aluminum cake pan

Electric stand mixer fitted with whisk attachment

Medium saucepan

Medium mixing bowl

Balloon whisk

Rubber or silicone spatula

Parchment paper

Toothpick

Cooling rack

Pizza cutter

Notes:

Plain Ole Brownies

This is an interesting mixing method that you'll often run into with very dense cakes and bars.

It starts like a curd and finishes with the Creaming Method.

THE DRY GOODS

INGREDIENT	Weight		Volume	Count	Prep
Cocoa powder	113 g	4 oz	1⅓ cups		
All-purpose flour	99 g	3½ oz	⅔ cup		
Kosher salt	3 g	<⅛oz	½ teaspoon		

THE WET WORKS

INGREDIENT	Weight		Volume	Count	Prep
Eggs	200 g	7 oz		4 large	
Vanilla extract	9 g	⅓ oz	2 teaspoons		
Sugar	198 g	7 oz	1 cup		sifted
Brown sugar	227 g	8 oz	1 cup		sifted
Unsalted butter	227 g	8 oz	1 cup	2 sticks	melted

THE EXTRAS

INGREDIENT	Weight		Volume	Count	Prep
Walnuts	85 g	3 oz	1 cup		
Baker's Joy or AB's Kustom Kitchen Lube for the pan					

✦ Place an oven rack in position C and preheat the oven to 350°F. Prepare an 8-inch square aluminum baking pan (see pages 180-183).

✦ Sift together the dry ingredients in the food processor.

✦ In an electric stand mixer fitted with a whisk attachment, whip the eggs at medium speed until light (both in texture and color). Add the vanilla.

◆ Mix the sugars together, reduce the mixer speed to 30-percent power, and add the sugars to the eggs, incorporating thoroughly.

◆ Add the batter and remaining dry ingredients in three alternating doses starting with the wet and finishing with the dry. Fold in the nuts.

◆ Pour the batter into the prepared pan and bake for 55-60 minutes. Check for doneness with the tried-and-true toothpick method: a toothpick inserted into the center of the pan should come out clean.

◆ Remove the pan to a cooling rack and resist the temptation to cut until the brownies are completely cool. When ready, cut into squares with a pizza cutter.

Yield: Sixteen 2-inch-square brownies

◆ ◆ ◆ ◆ ◆

Chocolate Frosting

INGREDIENT	Weight		Volume	Count	Prep
Bittersweet or semisweet chocolate	180g	6.5 oz	about 1 cup		
Whipping cream	114 g	4 oz	½ cup		
Unsalted butter	227 g	8 oz	1 cup	2 sticks	
Confectioners' sugar	285 g	10 oz	2½ cups		

THE EXTRAS

Ice to cool the frosting

◆ Combine the chocolate, cream, and butter in a saucepan and stir over medium heat until melted and smooth.

◆ Remove from heat and use a hand mixer to beat in powdered sugar. When dissolved, place pan inside a bowl filled with ice, then crank an electric hand mixer to medium and beat until frosting lightens and holds its shape.

Yield: 3½ cups

Hardware:

For the frosting
Digital scale
Dry measuring cups
Measuring spoons
Medium saucepan
Rubber or silicone spatula or wooden spoon
Large mixing bowl
Electric hand mixer
Offset spatula for frosting

Notes:

Store in refrigerator for up to a week.

Hardware:

Two 12-hole muffin tins

Digital scale

Dry measuring cups

Wet measuring cups

Measuring spoons

Food processor

Large mixing bowl

Stand mixer with paddle attachment

Rubber or silicone spatula

2½-ounce disher

Instant-read thermometer or toothpick

Cooling rack

Notes:

Bran Muffins

If you've read my *Gear* book, you've seen a earlier version of this recipe. I liked it then, but I like it more now…hey, that's how it is with recipes. Sure, some stay the same, but more often than not they evolve. Oh, and you'll no doubt notice that although these are muffins, they're not included in the Muffin Method section. That's because these devices possess a texture more like a cake. So while Carrot Cake is a muffin, Bran Muffins are cakes.

THE CREAMED

INGREDIENT	Weight		Volume	Count	Prep
Unsalted butter	113 g	4 oz	8 tablespoons	1 stick	softened
Dark brown sugar	170 g	6 oz	¾ cup		
Molasses	57 g	2 oz	¼ cup		

THE EGGS

INGREDIENT	Weight		Volume	Count	Prep
Eggs	100 g	3½ oz		2 large	

THE DRY GOODS

INGREDIENT	Weight		Volume	Count	Prep
Whole wheat flour	270 g	9½ oz	2 cups		
Baking soda	15 g	½ oz	2½ teaspoons		
Salt	6 g	<¼ oz	1 teaspoon		
Ground allspice	5 g	<¼ oz	1 teaspoon		

THE LIQUID

INGREDIENT	Weight		Volume	Count	Prep
Buttermilk	454 g	16 oz	2 cups		

THE EXTRAS

INGREDIENT	Weight		Volume	Count	Prep
Bran flakes	213 g	7½ oz	1 to 2 cups		
Toasted wheat bran or wheat germ	35 g	1¼ oz	½ cup		
Golden raisins	135 g	4¾ oz	1 cup		
Baker's Joy or AB's Kustom Kitchen Lube for the muffin tin					

✦ ✦ ✦ ✦ ✦

✦ Place an oven rack in position C and preheat the oven to 400°F.

✦ Prep the muffin tins (see pages 180–183) and set aside.

✦ Assemble the batter via the **CREAMING METHOD,** but when you're ready to add the Dry Goods, work in just half, then alternate with half of the buttermilk, then the remaining flour mixture, then the rest of buttermilk. Fold in the bran flakes, toasted wheat bran, and raisins and stir just until combined.

✦ Using the disher, spoon the batter into the tins, filling them to the top (about 2½ to 3 ounces).

✦ Bake for 20 minutes, until the muffins reach an internal temperature of 210°F or a toothpick inserted into the center of a muffin comes out clean.

✦ Allow the muffins to cool before taking them out of the tins. They'll keep in an air-tight container for up to 5 days.

Yield: About 20 standard muffins

Note: This batter keeps in the fridge for up to 2 weeks. Allow an additional 5 to 7 minutes in the oven when you're using refrigerated batter.

Notes:

Hardware:

12-hole muffin tin

Digital scale

Dry measuring cups

Wet measuring cups

Measuring spoons

Chef's knife

Cutting board

Juicer

Zester or microplane grater

Food processor

Large mixing bowl

Stand mixer with paddle attachment

Rubber or silicone spatula

2½-ounce disher

Instant-read thermometer or toothpick

Notes:

Or you can remove the orange zest (and only the zest, not the pith) with a peeler and mince it fine with a chef's knife.

Orange Cranberry Muffins

This application is interesting because it's a hybrid—at first the batter goes together via creaming, then it finishes like a muffin. The result is not as rustic as an old-school muffin but not as refined as a cupcake.

THE CREAMED

INGREDIENT	Weight		Volume	Count	Prep
Unsalted butter	113 g	4 oz	8 tablespoons	1 stick	softened
Sugar	198 g	7 oz	1 cup		

THE EGGS

INGREDIENT	Weight		Volume	Count	Prep
Eggs	100 g	3½ oz		2 large	beaten

THE DRY GOODS

INGREDIENT	Weight		Volume	Count	Prep
All-purpose flour	270 g	9½ oz	2 cups		
Baking powder	3 g	<⅛ oz	½ teaspoon		
Baking soda	6 g	<¼ oz	1 teaspoon		
Salt	3 g	<⅛ oz	½ teaspoon		

THE LIQUID

INGREDIENT	Weight		Volume	Count	Prep
Orange extract	5 g	⅙ oz	1 teaspoon		
Plain yogurt	112 g	4 oz	½ cup		
Fresh orange juice	43 g	1½ oz	1½ ounces	1 orange	
Orange zest	11 g	⅜ oz	1 tablespoon	1 orange	
Canned whole-berry cranberry sauce	170 g	6 oz	about ¾ cup		

THE EXTRAS

One recipe Streusel (see page 160) ✦ Baker's Joy or AB's Kustom Kitchen Lube for the muffin tins

Notes:

✦ ✦ ✦ ✦ ✦

✦ Make the streusel according to the recipe on page 160 and set aside.

✦ Prep the muffin tin (see pages 180–183) and set aside.

✦ Place an oven rack in position C and preheat the oven to 350°F.

✦ Combine the Dry Goods via the food processor and place in a large mixing bowl.

✦ Assemble the batter via the **CREAMING METHOD** through Step 5, the integration of the Eggs into the Creamed.

✦ Finish the batter via the **MUFFIN METHOD,** that is, combine the remaining Liquids and add them to the Creamed. Then dump all the wet ingredients onto the Dry Goods and stir until the batter just comes together.

✦ Using the disher, scoop the batter into the tins, filling them to the top (2½ to 3 ounces). Top with the Streusel and bake for 30 to 35 minutes, until the interior reaches a temperature of 210°F, or a toothpick inserted into the center of a muffin comes out clean.

✦ Allow the muffins to cool before taking them out of the tins. These will keep, in an air-tight container, for up to 5 days.

Yield: 12 standard muffins

Hardware:

Digital scale

Dry measuring cups

Wet measuring cups

Measuring spoons

Chef's knife

Cutting board

Food processor

2 large mixing bowls

Stand mixer with paddle attachment

Rubber or silicone spatula

Wax paper

2 cookie sheets

Metal spatula

Cooling rack

Notes:

Ginger Cookies

Like a gingersnap, only chewier. Note that you will refrigerate the dough overnight before baking.

THE CREAMED

INGREDIENT	Weight		Volume	Count	Prep
Unsalted butter	113 g	4 oz	8 tablespoons	1 stick	softened
Dark brown sugar	227 g	8 oz	1 cup		
Molasses			1 tablespoon		

THE EGGS

INGREDIENT	Weight		Volume	Count	Prep
Eggs	50 g	1¾ oz		1 large	

THE DRY GOODS

INGREDIENT	Weight		Volume	Count	Prep
All-purpose flour	270 g	9½ oz	2 cups		
Baking powder	2 g	<⅛ oz	½ teaspoon		
Baking soda	3 g	<⅛ oz	½ teaspoon		
Ground ginger	12 g	<½ oz	2 teaspoons		
Salt	2 g	<⅛ oz	¼ teaspoon		

THE EXTRAS

INGREDIENT	Weight		Volume	Count	Prep
Crystallized (candied) ginger			¼ cup		chopped to a small dice
Parchment paper for the cookie sheets					

◆ ◆ ◆ ◆ ◆

✦ Assemble the dough via the **CREAMING METHOD** and stir in the crystallized ginger.

✦ Roll the dough into two logs, 2 inches in diameter. Wrap them in wax paper, and refrigerate overnight.

✦ The next day, place two oven racks in positions B and C and preheat the oven to 375°F.

✦ Prep the cookie sheets (see pages 180–183) and set aside.

✦ Remove the dough from the refrigerator and cut each log into ¼-inch slices.

✦ Place the slices on the prepared cookie sheets, and bake for 8 to 10 minutes, until the cookies have soft set, rotating the pans after 4 minutes.

✦ Remove the pans from the oven and move the cookies onto a rack to cool.

✦ Store in an air-tight container for up to 2 weeks.

Yield: Three dozen 2-inch cookies

> Meaning that the centers of the cookies have set, but the cookies are still soft.

> I move the pans from one rack to the other and rotate the trays front to back.

> Yield accuracy goes downhill if you eat the batter straight from the bowl.

ABOUT COOKIE PORTIONS

◆ ◆ ◆

Once upon a time you could count on the numbers etched into the side or the sweeper of dishers. I love dishers and have about ten in different sizes. The problem is, you can't trust them anymore. See, the number that's etched on the side or on the sweeper itself is supposed to state how many scoops will make a quart—so a #32 is a 1-ounce portion. But not any more. These days you can find three different #20s, all different sizes. So, in order to come up with accurate yields we're going with 1-ounce cookies across the board. That means that you'll have to weigh a few until you get the hang of what an ounce looks and feels like. When rolled into a rough sphere, it'll range between a ping-pong ball and a golf ball, depending on the dough in question. If you think that a 1-ounce cookie is stingy, then just eat more cookies.

While we're on the subject of yields, you may notice that the cookie recipes in this book, by and large, make a heap of cookies. If you don't really want to bake that many, I suggest you make and scoop the entire batch, then place however many portions you don't want to bake on a parchment-lined sheet pan and park them in the freezer overnight. When they're hard as a rock, bag 'em in zip-top freezer bags and freeze for up to 6 months. When you're ready to bake, move the blobs to a sheet pan before you turn on the oven. By the time the oven hits its target temp, the cookies will be thawed enough to bake. Or you can simply eat them like ice cream. Yes, I sometimes do this.

Hardware:

2 cookie sheets

Digital scale

Dry measuring cups

Wet measuring cups

Measuring spoons

Food processor

2 large mixing bowls

Stand mixer with paddle attachment

1-ounce disher

Kitchen fork

Metal spatula

Cooling rack

Notes:

Peanut Butter Cookies

This is one of those types of cookies that people have strong feelings about. I know I can't please everyone, but this comes pretty darned close to pleasing me.

THE CREAMED

INGREDIENT	Weight		Volume	Count	Prep
Unsalted butter	340 g	12 oz	1½ cups	3 sticks	softened
Granulated sugar	284 g	10 oz	1½ cups		
Brown sugar	284 g	10 oz	1¾ cups		
Peanut or canola oil	78 g	2¾ oz	½ cup		
Chunky peanut butter	567 g	20¾ oz	2 cups + 1 tablespoon		measure by weight, please

THE EGGS

INGREDIENT	Weight		Volume	Count	Prep
Eggs	150 g	5¼ oz		3 large	beaten
Vanilla extract	9 g	⅓ oz	2 teaspoons		

THE DRY GOODS

INGREDIENT	Weight		Volume	Count	Prep
All-purpose flour	510 g	18 oz	3¾ cups		
Baking soda	12 g	½ oz	2 teaspoons		
Salt	9 g	<⅓ oz	1½ teaspoons		

THE EXTRAS

INGREDIENT	Weight		Volume	Count	Prep
Granulated sugar for sprinkling on top	50 g	1¾ oz	¼ cup		
Parchment paper for the cookie sheet					

Please, in the name of all that is good and pure, weigh this stuff—especially this stuff.

✦   ✦   ✦   ✦   ✦

✦ Place oven racks in positions B and C and preheat the oven to 350°F.

✦ Prep the cookie sheets (see pages 180–183) and set aside.

✦ Assemble the Dry Goods, then combine the butter and sugars via the **CREAMING METHOD** through Step 4.

✦ When the butter-sugar mixture has lightened, drop the mixer speed to Low and add the oil and peanut butter in a single dose. Increase the mixer speed to 50 percent and cream for another 2 minutes until well combined.

✦ Add the Liquid and finish mixing the dough via the **CREAMING METHOD.**

✦ Place the dough in the refrigerator and chill for a half hour, then—using the disher—portion into golf ball–size balls and place them on the cookie sheets. Use a fork to press the dough down in two directions to create a criss-cross pattern.

✦ Sprinkle the cookie tops with the additional granulated sugar.

✦ Bake for 17 to 20 minutes, rotating the pans after 10 minutes, or until the cookies are lightly browned around the edges.

✦ Remove from the oven and let the cookies sit for 2 minutes on the pans before removing them to a rack to cool completely.

✦ The cookies will keep in an air-tight container for 2 weeks.

Yield: About 6 dozen cookies

Notes:

You'll definitely have to scrape down the side of the bowl a couple of times.

Oatmeal Cookies

This recipe calls for plain ole regular oatmeal. Don't use quick, instant, or any of the fancy Irish stuff, which I do like, but not for cookies.

THE CREAMED

INGREDIENT	Weight		Volume	Count	Prep
Unsalted butter	567 g	20 oz	2½ cups	5 sticks	softened
Dark brown sugar	340 g	12 oz	1½ cups		
Sugar	198 g	7 oz	1 cup		

THE EGGS

INGREDIENT	Weight		Volume	Count	Prep
Eggs	100 g	3½ oz		2 large	
Vanilla extract	9 g	⅓ oz	2 teaspoons		

THE DRY GOODS

INGREDIENT	Weight		Volume	Count	Prep
All-purpose flour	411 g	14½ oz	3 cups		
Baking powder	7 g	¼ oz	2 teaspoons		
Ground cinnamon	7 g	<¼ oz	2 teaspoons		

THE EXTRAS

INGREDIENT	Weight		Volume	Count	Prep
Rolled oats	468 g	16¾ oz	6 cups		
Raisins	227 g	8 oz	2 cups		
Parchment paper for the cookie sheets					

Hardware:

Digital scale

Dry measuring cups

Wet measuring cups

Measuring spoons

Food processor

2 large mixing bowls

Stand mixer with paddle attachment

1-ounce disher

Parchment paper, if using

2 cookie sheets

Metal spatula

Cooling rack

Notes:

Notes:

✦ ✦ ✦ ✦ ✦

✦ Place oven racks in positions B and C and preheat the oven to 375°F.

✦ Assemble the dough via the **CREAMING METHOD,** stirring the oats and the raisins into the mix last.

✦ Using the disher, scoop and drop the dough onto ungreased or parchment-lined cookie sheets, with each mound of dough at least 2 inches apart.

✦ Bake for 15 to 17 minutes, rotating the pans after 8 minutes, until the cookies have begun to brown at the edges.

✦ Remove from oven and hold the cookies on the baking sheets for 2 minutes before moving them to a rack to cool completely.

✦ The cookies will keep in an air-tight container for 2 weeks.

Yield: About 6 dozen cookies

Hardware:

9 x 5 x 2¾-inch nonstick loaf pan

Digital scale

Dry measuring cups

Wet measuring cups

Measuring spoons

Nutmeg grater or microplane

Chef's knife

Cutting board

Food processor

Large mixing bowl

Small mixing bowl

Whisk

Stand mixer with paddle attachment

Rubber or silicone spatula

Instant-read thermometer or toothpick

Cooling rack

Notes:

Dried Fig Hazelnut Bread

In terms of method, halfway between a banana bread and fruitcake, this application features another hybrid procedure beginning with the **CREAMING METHOD** and finishing with the **MUFFIN METHOD**.

THE FIGS

INGREDIENT	Weight		Volume	Count	Prep
Dried whole figs	170 g	6 oz	approximately 1¼ cups		
Boiling water	227 g	8 oz	1 cup		

THE CREAMED

INGREDIENT	Weight		Volume	Count	Prep
Unsalted butter	85 g	3 oz	6 tablespoons		softened
Dark brown sugar	170 g	6 oz	¾ cup		

THE EGGS

INGREDIENT	Weight		Volume	Count	Prep
Eggs	100 g	3½ oz		2 large	

THE DRY GOODS

INGREDIENT	Weight		Volume	Count	Prep
All-purpose flour	270 g	9½ oz	2 cups		
Baking powder	7 g	¼ oz	2 teaspoons		
Salt	3 g	<⅛ oz	½ teaspoon		
Freshly grated nutmeg	4 g	<¼ oz	1 teaspoon		
Ground ginger	2 g	<⅛ oz	½ teaspoon		

THE LIQUID

INGREDIENT	Weight		Volume	Count	Prep
Whole milk	170 g	6 oz	¾ cup		

Notes:

THE EXTRAS

INGREDIENT	Weight		Volume	Count	Prep
Hazelnuts	113 g	4 oz	¾ cup		roasted and coarsely chopped
Shortening and parchment paper for the pan					

✦ ✦ ✦ ✦ ✦

✦ Soak the figs in the boiling water for 20 to 30 minutes until softened.

✦ Place an oven rack in position C and preheat the oven to 375°F.

✦ Prep the loaf pan (see pages 180–183) and set aside.

✦ Assemble the Dry Goods, the Creamed, and the Eggs via the **CREAMING METHOD** through Step 5.

✦ Add the milk to the creamed ingredients.

✦ Drain the figs and chop them into a small dice.

✦ Apply the **MUFFIN METHOD** by adding the wet ingredients to the dry and stirring just until combined.

✦ Fold in the figs and hazelnuts. Pour the batter into the prepared loaf pan and bake for 50 to 60 minutes, until the internal temperature reaches 210°F, or a toothpick inserted in the center of the loaf comes out clean.

✦ Remove the bread from the pan and place on a rack to cool before slicing.

✦ Wrapped and refrigerated, this will keep for a month. You can also freeze it for up to 3 months.

Yield: One 9-inch loaf

Pop Goes the Tart

The toaster, developed at the dawn of the twentieth century, is probably America's greatest contribution to the culinary world. As the first electric kitchen appliance to be mass marketed (and mass purchased), the toaster heralded a new era.

When I was a kid I had an affinity for prepackaged pastries filled with fruity goodness and designed to slide down the gullet of a toaster, which just happened to be the one heat-emitting culinary device my mom allowed me to use. I consumed thousands of these deadly devices as a child, but when my big boy teeth came in I dispensed with such childishness. But then about a year ago I started thinking that those tidy rectangles might be just the thing to snap me out of the monotony of my grab-n-go breakfast malaise. So I picked some up and…they tasted horrible. Clearly, my tastes had changed. But I didn't want to give up on the treat I had once enjoyed—so I decided to make them myself. After much trial-and-error, here's the result—and a fine result it is.

More and you're likely to pop your tart from overfilling.

THE CREAMED

INGREDIENT	Weight		Volume	Count	Prep
Sugar	227 g	8 oz	1 cup + 2 tablespoons		
Vegetable shortening	184 g	6½ oz	1 cup		70°F

THE EGGS

INGREDIENT	Weight		Volume	Count	Prep
Eggs	100 g	3½ oz		2 large	

THE DRY GOODS

INGREDIENT	Weight		Volume	Count	Prep
All-purpose flour	595 g	21¼ oz	4½ cups		
Baking powder	7 g	¼ oz	2 teaspoons		
Salt	3 g	< ⅛ oz	½ teaspoon		

THE LIQUID

INGREDIENT	Weight		Volume	Count	Prep
Milk	57 g	2 oz	¼ cup		

THE EXTRAS

INGREDIENT	Weight		Volume	Count	Prep
Jam, fruit butter, or preserves	595 g	21 oz	½ cup		
Egg wash (glue)	50 g	1¾ oz		1 large egg	
Milk	28 g	1 oz	2 tablespoons		

Hardware:

Digital scale

Dry measuring cups

Wet measuring cups

Measuring spoons

Food processor

Large mixing bowl

2 small mixing bowls

Whisk

Stand mixer with paddle attachment

Rubber or silicone spatula

Wax paper

2 sheet pans

Rolling pin (preferably the French kind)

Bench scraper

Kitchen fork

Cooling rack

✦ ✦ ✦ ✦ ✦

✦ Assemble the dough via the **CREAMING METHOD,** alternating 3 doses of the dry mixture with 2 doses of milk.

✦ Place oven racks in positions B and C and preheat the oven to 350°F.

✦ Get out several of your favorite jams, preserves, and butters—as in apple butter.

1 Place the dough onto a sheet of waxed paper—about 13 x 17 inches long—and fold the end of the paper over.

2 Using a sheet pan as a wedge, continue to roll the dough into a log. This procedure tightens the log into a clean cylinder. Remove the pan lip, finish the roll, and then fold in the ends.

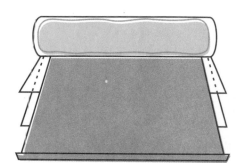

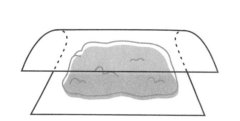

3 Remove the waxed paper, slice the dough into 2-inch rounds (you should have 16 pieces), place on a sheet pan, cover and chill for 1 hour.

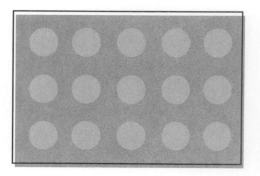

As is true of just about every pastry dough on the planet, the cooler the work environment the easier the task. I've got a screened-in porch with a wooden table, so on cool evenings I roll out there. It's sorta peaceful like.

✦ When the dough is chilled, get ready to roll.

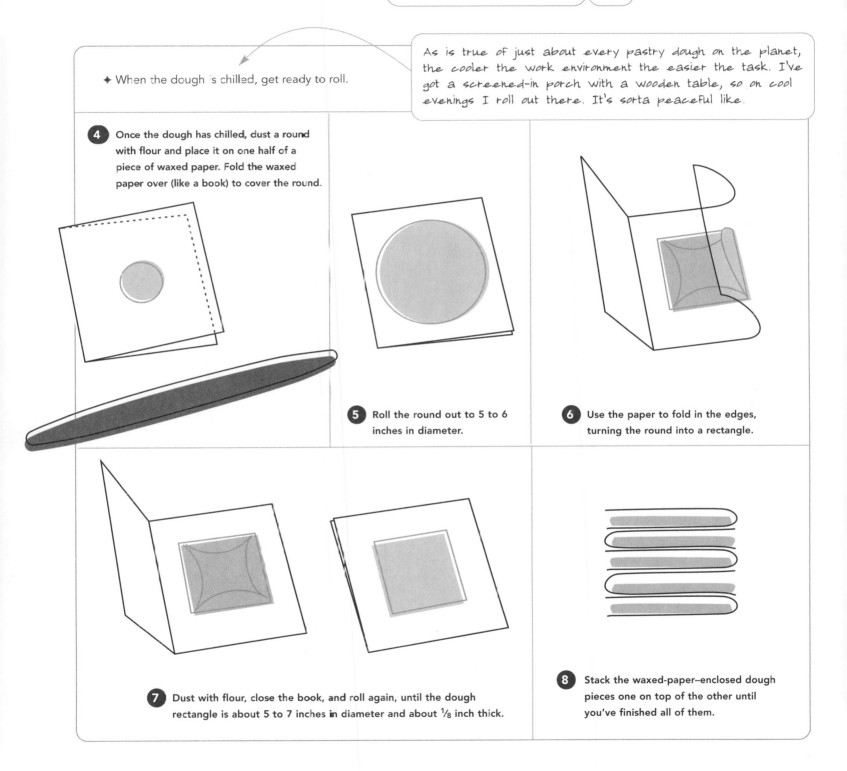

4 Once the dough has chilled, dust a round with flour and place it on one half of a piece of waxed paper. Fold the waxed paper over (like a book) to cover the round.

5 Roll the round out to 5 to 6 inches in diameter.

6 Use the paper to fold in the edges, turning the round into a rectangle.

7 Dust with flour, close the book, and roll again, until the dough rectangle is about 5 to 7 inches in diameter and about ⅛ inch thick.

8 Stack the waxed-paper–enclosed dough pieces one on top of the other until you've finished all of them.

Notes:

✦ Take a look at your pieces and try to pair them up. Sure, in a perfect world they'd all be exactly alike. This is not a perfect world, so try to match up good pairs.

✦ Mix up your egg wash—actually, you should have done that while the dough was chilling. What's the matter with you anyway?

✦ Now follow the pictures:

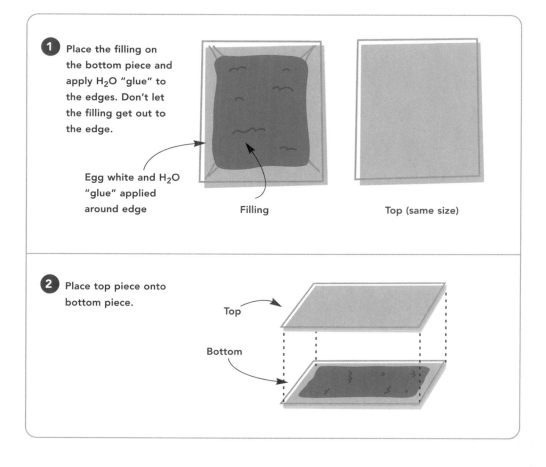

1 Place the filling on the bottom piece and apply H_2O "glue" to the edges. Don't let the filling get out to the edge.

Egg white and H_2O "glue" applied around edge

Filling

Top (same size)

2 Place top piece onto bottom piece.

Top

Bottom

✦ The forking, or "docking," is crucial. If you skip this step, the water inside the preserves will blow the tarts up like balloons. But beware—if you poke with too much gusto, you might go through the bottom piece of dough, and preserves will ooze out the bottom where the sugar will fuse to the pan. Chisels will be required to remove the baked-on mess.

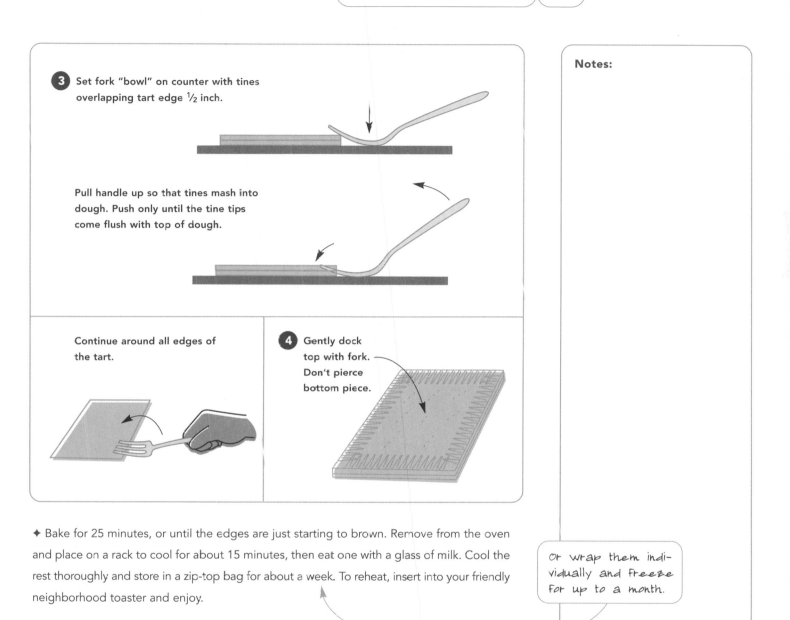

3 Set fork "bowl" on counter with tines overlapping tart edge ¹/₂ inch.

Pull handle up so that tines mash into dough. Push only until the tine tips come flush with top of dough.

Continue around all edges of the tart.

4 Gently dock top with fork. Don't pierce bottom piece.

Notes:

✦ Bake for 25 minutes, or until the edges are just starting to brown. Remove from the oven and place on a rack to cool for about 15 minutes, then eat one with a glass of milk. Cool the rest thoroughly and store in a zip-top bag for about a week. To reheat, insert into your friendly neighborhood toaster and enjoy.

Or wrap them individually and freeze for up to a month.

Yield: 8 tarts

Note: Yes, you could frost these, but I really don't think that…oh, okay. In a small bowl slowly whisk 4 ounces of confectioners' sugar into 2 tablespoons whole milk, then paint on to the front of the tarts. Allow to dry before stacking, storing, or snarfing.

Hardware:

10-inch aluminum tube pan

Digital scale

Dry measuring cups

Wet measuring cups

Measuring spoons

Food processor

Large mixing bowl

Small mixing bowl

Whisk

Stand mixer with paddle attachment

Rubber or silicone spatula

Cooling rack

Notes:

Sweet milk is simply regular whole milk. In the days when people churned their own butter, buttermilk (the natural residue of the churned butter) was much more common on this country's tables. In fact, it was so common that many recipes—especially Southern recipes—distinguished whole milk from buttermilk by referring to it as "sweet."

Chocolate Pound Cake

I have no idea where this recipe came from, but since it calls for "sweet" milk, I have to think that it's at least forty years old, if not older. If you want to kick up the chocolate "a few notches," then consider making a slurry of the milk and cocoa powder and bringing it to a boil in the microwave (a couple of minutes does it in mine). The boiling liquid "opens up" the chocolate, giving you more bang for your nib, so to speak. That said, this cake is just dandy the way it is described below.

THE CREAMED

INGREDIENT	Weight		Volume	Count	Prep
Unsalted butter	227 g	8 oz	1 cup	2 sticks	softened
Sugar	595 g	21 oz	3 cups		

THE EGGS

INGREDIENT	Weight		Volume	Count	Prep
Eggs	250 g	8¾ oz		5 large	beaten
Vanilla extract	9 g	⅓ oz	2 teaspoons		

THE DRY GOODS

INGREDIENT	Weight		Volume	Count	Prep
All-purpose flour	404 g	14½ oz	3 cups		sifted once, then twice more with the cocoa
Cocoa powder	18 g	⅝ oz	6 tablespoons		
Baking powder	2 g	<⅛ oz	½ teaspoon		
Salt	2 g	<⅛ oz	⅛ teaspoon		

THE LIQUID

INGREDIENT	Weight		Volume	Count	Prep
Sweet milk	227 g	8 oz	1 cup		

THE EXTRAS

Baker's Joy or AB's Kustom Kitchen Lube for the pan

✦   ✦   ✦   ✦   ✦

✦ Place an oven rack in position B and preheat the oven to 325°F.

✦ Prep a 10-inch tube pan (see pages 180–133) and set aside.

✦ Assemble the batter via the **CREAMING METHOD**, alternating additions of the flour/cocoa mixture with the milk. Given the amounts involved, start with dry and end with dry.

✦ Pour the batter into the pan and bake for 1 hour and 20 minutes, until the internal temperature hits 212°F, or the cake leaves the sides of the pan.

✦ Remove the cake from the oven and allow to cool 15 minutes in the pan, then turn out onto a rack to cool thoroughly.

✦ It will keep, tightly wrapped at room temperature, for 1 week.

Yield: One 10-inch cake

Hardware:

12-cup tube or Bundt pan

Digital scale

Dry measuring cups

Wet measuring cups

Measuring spoons

Food processor

Large mixing bowl

Small mixing bowl

Whisk

Stand mixer with paddle attachment

Rubber or silicone spatula

Cooling rack

Notes:

Pound Cake

It's the buttermilk that makes this so gosh-darned good. All that acid gets with the soda and takes care of business. If you've never ever made a cake, this is a good place to start. It's good toasted for breakfast, too, and leftovers make for a very nice bread pudding…if it lasts that long.

THE CREAMED

INGREDIENT	Weight		Volume	Count	Prep
Unsalted butter	227 g	8 oz	1 cup	2 sticks	softened
Sugar	397 g	14 oz	2 cups		

THE EGGS

INGREDIENT	Weight		Volume	Count	Prep
Eggs	150 g	5¼ oz		3 large	beaten
Vanilla extract	5 g	<¼ oz	1 teaspoon		

THE DRY GOODS

INGREDIENT	Weight		Volume	Count	Prep
All-purpose flour	411 g	14½ oz	3 cups		
Baking soda	3 g	⅛ oz	½ teaspoon		
Salt	3 g	<⅛ oz	½ teaspoon		

THE LIQUID

INGREDIENT	Weight		Volume	Count	Prep
Buttermilk	227 g	8 oz	1 cup		full fat, if possible

THE EXTRAS

Baker's Joy or AB's Kustom Kitchen Lube for the pan

Notes:

✦ ✦ ✦ ✦ ✦

✦ Place an oven rack in position B and preheat the oven to 325°F.

✦ Prep a tube or Bundt pan (see pages 180–183) and set aside.

✦ Assemble the batter via the **CREAMING METHOD,** alternating additions of the Dry Goods with the Liquid.

✦ Pour the batter into the pan and bake for 1 hour, or until the cake pulls away from the sides of the pan and the internal temperature hits 212°F.

✦ Remove from the oven and allow to cool 15 minutes in the pan, then turn out onto a rack to cool thoroughly. Tightly wrapped at room temp, the cake will keep for 1 week.

Yield: Serves 6 to 8

The Straight Dough Method

The mission here is to create a dough with the plasticity and elasticity necessary to accommodate the gas produced by living yeast (think of well-chewed bubble gum). When baked, the texture of the resulting bread may be tight and even or open and irregular, depending on how much moisture is present, how much the dough was kneaded, and the time and temperature at which the yeast had to do their work.

The Straight Dough Method

Assumes the use of instant yeast and kosher salt.

1. Properly scale/measure ingredients.

2. Mix yeast and salt with all but 1 cup of the flour (I do this with a food processor).

3. Add liquids and sugar (if called for) to work bowl, then add flour-yeast mixture.

4. Mix to thoroughly combine.

5. Cover and rest 20 to 30 minutes.

6. Knead, adding remaining flour a little at a time.

7. Cover and allow dough to rise until doubled in volume.

8. Move to lightly floured surface and redistribute yeast and gas by flattening and folding dough.

9. Rest dough for 5 minutes to relax gluten, then form into final shape or allow dough to rise again.

10. Cover for second (or third) rise until doubled in size.

11. Slash and bake according to the recipe.

12. Cool thoroughly before slicing.

The Sponge Method Variation

Follow the procedure outlined above, with the following exceptions:

2. Use only half of the prescribed flour. Do not add salt during this step.

6. Add the salt with any remaining flour.

The Straight Dough Method

+ If more than one type of flour is called for, mix them together before adding them to the rest of the ingredients.

+ For the first fermentation (step 5 on the other side of this flap), take a heating pad, turn it up to high, set it inside a mixing bowl, and set the bowl containing the dough on top of that. The warmth will speed the fermentation.

Although there are certainly more ways to build a yeast dough, I'd be willing to bet that more than 90 percent of the yeast-bearing applications on the planet are constructed using the traditional Straight Dough Method.

It usually goes something like this:

1. Scale or measure ingredients.

2. Soak yeast in warm water (with small portion of sugar, if using).

3. Add ingredients to work bowl, starting with liquids, followed by dry ingredients, and, finally, salt.

4. Mix into a dough.

5. Turn out onto a board or other floured surface and knead until smooth and fully developed.

6. Cover and allow to rise until doubled.

7. Punch down, knead briefly.

8. Rise again, or shape into final form.

9. Bench proof.

10. Slash and bake.

11. Cool.

Pretty simple, huh? Let's take a leisurely look at each step, shall we?
Scale or measure ingredients. You often hear bakers say that baking is an exact science. It's not. It's an exacting craft. By that I mean that almost all the factors involved can be fudged, shifted, changed—within reason, that is, *and* only if you know what you're

doing. For instance, the amount of flour that actually goes into the dough for a given bread recipe may never be the same twice. What's the weather like (humid, dry)? What's the temperature in the kitchen? What kind of flour is being used? There are just too many variables for the science of baking to be "exact." And on any given day, the cook may want a bread with a finer texture, or a softer, more open one—all that can be affected to some extent by how much flour is coaxed into a recipe.

All that said, I still weigh just about everything that goes into a bread, including the water, because it's just easier. And at home I always work with metric measurements because grams are a lot easier to deal with than fractions and decimal points.

Soak yeast in warm water. Most bread recipes written between World War II and the late 1980s include this step. Today's "instant" yeasts don't require such treatment (see pages 72–77 for more yeast chat). If you're forced by circumstances beyond your control to use active dry yeast, then sprinkle the yeast over ¼ cup or so of the water—heated to between 105 and 115°F—which you'll be using for the recipe. The general rule is that yeast needs to soak in four times its weight in water, but I usually just go with the ¼ cup (2 ounces by weight or volume)—because it's an easy amount to deal with. Then add this to the rest of the ingredients as described below.

Add ingredients to a work bowl. Since I'm never sure how much flour really is going to fit into a dough on any particular day and since my yeast aren't floating around in water, I usually set the last cup of flour to the side and sift the flour and the instant yeast together in the food processor before adding them to the work bowl. Processing the two together will ensure even distribution of the yeast and that's the most important part of the mixing procedure. Despite what you might have heard, you *can* add instant yeast and salt to the mixture at the same time—unless you're using the sponge variation discussed later in this section.

Mix into a dough. If you want to use an electric stand mixer for this, the dough hook is not effective, so you'll need the paddle attachment. I hate cleaning that thing, so I usually use a big spatula or (gasp!) my hands. Whatever you use, the point of the mix step is just that: to bring everything together.

Once the dough is mixed together but before it's kneaded, cover the bowl with a towel and let it rest for half an hour. Professional bakers have long known that this rest, which is called autolyse (*auto-lease*), gives the starches and proteins in the flour time to hydrate. That will make the dough a lot easier to knead and will result in better gluten structure (and good gluten structure is a good thing). Recipe writers have known about this step for a long time, too, but have been hesitant to add another time-consuming step to a process for which a majority of Americans already have no patience.

Kneading refers to the mechanical stretching and folding of the dough—not punching, poking, or jabbing. Successful kneading by hand requires leaning into the dough mass, pushing it away, stretching it, and then folding it over—and then repeating these steps over and over again. Although it is possible to over-knead dough, it's not very likely that you'll do so. Most people under-knead or "underdevelop" when working by hand. What happens is they just get tired. If you do manage to over-knead, you'll know it because suddenly the dough will collapse and water will run out as if you'd wrung out a wet washcloth. I've only seen this happen once, when someone (who looked a lot like me) abandoned an industrial mixer full of pumpernickel…or maybe it was rye.

I also add my final dose of flour to the dough during kneading. Since people tend to flour the board and their hands in order to handle the dough, they often end up working in more than the dough needs. The result is a dry dough that doesn't rise well because there's not enough water left for the yeast to work in—it's all been soaked up by the flour. These loaves don't "spring" well in the oven either because there isn't enough free water to turn to steam.

The words **smooth** and **developed** pertain to the texture of the dough once the gluten formation is in place. How do you know when you've achieved this? Well, bread dough is like bubble gum. When you first pop a fresh hunk of bubble gum in your mouth, you can't immediately blow a bubble, can you? Of course not. First you have to do some serious chewing. The chewing breaks up the gum, while your saliva dissolves the sugar and lubricates the gum's structure so that it can stretch. After a few minutes of jaw work, you're free to finagle the wad into a disk. Then, using your tongue, you turn that disk into

Kneading by hand

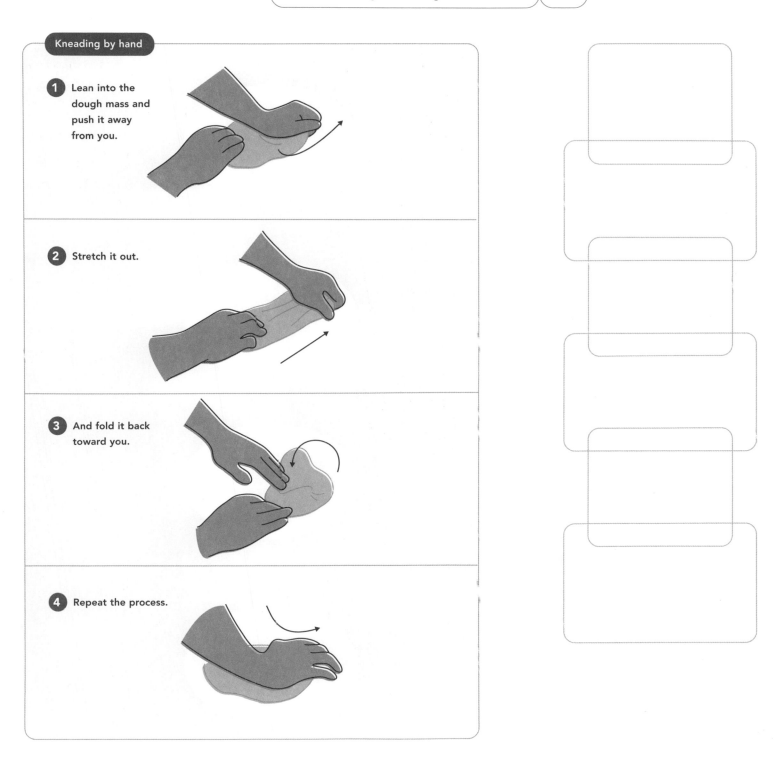

1 Lean into the dough mass and push it away from you.

2 Stretch it out.

3 And fold it back toward you.

4 Repeat the process.

an inflatable pocket. Kneading bread dough is just like chewing that gum so that it can form thin membranes that can house bubbles. There, you see—everything in baking is about bubbles.

So how do you know when the dough is fully "developed"? Most bakers will tell you that it's just experience, and they'll be right. But there is a little test you can do that will tell you a lot. I call it "windowpaning" and, although it's not the easiest thing to explain in words, we'll try to add a few pictures to the mix:

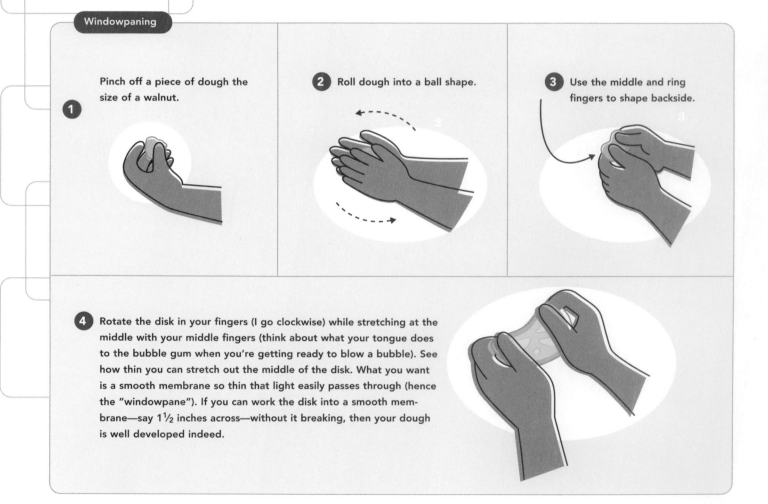

Windowpaning

1 Pinch off a piece of dough the size of a walnut.

2 Roll dough into a ball shape.

3 Use the middle and ring fingers to shape backside.

4 Rotate the disk in your fingers (I go clockwise) while stretching at the middle with your middle fingers (think about what your tongue does to the bubble gum when you're getting ready to blow a bubble). See how thin you can stretch out the middle of the disk. What you want is a smooth membrane so thin that light easily passes through (hence the "windowpane"). If you can work the disk into a smooth membrane—say 1½ inches across—without it breaking, then your dough is well developed indeed.

Rising. Just about every yeast-dough recipe on the planet requires that the dough increase in volume by a specific amount during the first rise (officially known as the "primary fermentation"). Although time guidelines are given, the real indicator of how the wee beasties (yeasts) are doing is the volume.

Since this is almost impossible to gauge in a bowl, I do my risin' in polycarbonate (clear, hard plastic) food storage containers. I have two—one for small batches and another for large—which I picked up at a local restaurant supply house. Both are made by Carlisle, a food service supply manufacturer. Tupperware and Rubbermaid make similarly styled vessels, but only food service–oriented stuff will get you into the 2-to-5-gallon range you'll need for bigger batches. Here's how it works for me.

The small (1 gallon) container is round and relatively narrow. I've placed two wide rubber bands around it, one that's red and another that's blue. Before I add the newborn dough, I spray the entire inside with nonstick spray and I let it air out a bit.

Most nonstick sprays contain a good bit of alcohol, and I'd just as soon that dissipate before the dough goes in.

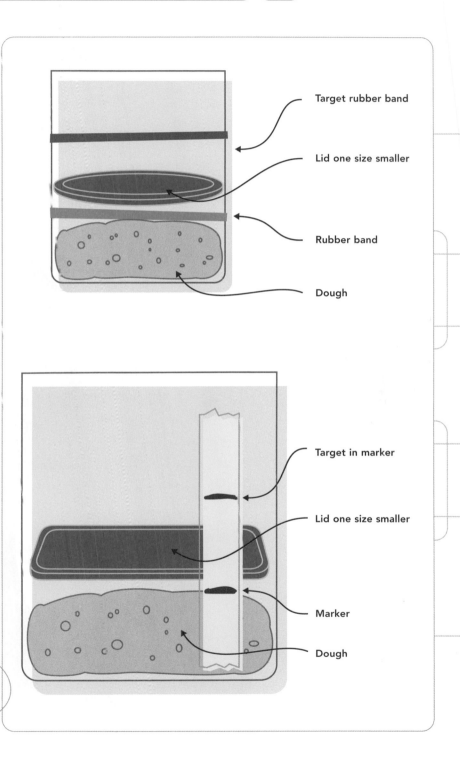

Target rubber band

Lid one size smaller

Rubber band

Dough

Target in marker

Lid one size smaller

Marker

Dough

I spray the underside of the lid with nonstick, too.

The dough goes in and I spread it around so that it's not just a wad. Then I top it with the lid from the container that's the *next size down*. This way the lid fits down inside the container and rests right on the surface of the dough to keep it from drying out.

Before I walk away I position the blue rubber band so that it is even with the top of the dough. I then position the red rubber band so that it is twice as high up the side. That way, when the dough reaches the red rubber band I know that it's doubled in volume. Could I do with just one rubber band? Sure, but using two makes the system a little more precise.

Since the vessel I use for really big loads is too big around for any rubber bands I've got lying around, I just run a piece of masking tape up the side and mark it with a magic marker. I suppose you could tape a ruler to the side for even more precise measurements …or now that I think of it, just use an permanent marker to index the side of the vessel in inches (or centimeters, if you prefer). Either way, you get the point: *Use a relatively narrow container and an indexing system of some type to measure dough rise.*

How long this takes depends to a great extent on the temperature of the dough. Although a lot of recipes call for rising in a "warm place," I like a long, cool rise. That allows the dough to absorb more of the flavor produced by the yeast. With a fast rise, you just end up with a lot of gas, and while that's an impressive display of fungal metabolism, most of it is released when the dough is "punched down."

Any yeast dough can get the slow-rise treatment. It will *always* result in better flavor and texture. But if you don't have time for the slow rise, remember that even fast-fermented homemade bread beats the pants off of store-bought. In either case, the general guideline is to allow the dough to double in volume. Depending on the dough, this could take as little as an hour at room temperature or as long as eight hours in the fridge.

There are many other factors that affect fermentation, including room temperature, beginning dough temp, type of yeast used, fat content, sugar content, and so on. Generally, fat and sugar slow fermentation.

Punching down. This is one of the most damaging misnomers in baking, created, I suspect, by sneaky professional bakers who didn't want anyone to be able to produce decent bread at home. Punching dough with your fists is a bad thing. One, it toughens the structure of the dough and, two, your hand gets stuck in it like Brer Rabbit in the Tar Baby.

The goal of this step isn't to simply knock out all the gas created by the yeast during the first fermentation, it's to evenly distribute small bubbles of gas and—above all—to distribute yeast cells, which tend to clump up when they reproduce. Why?

Since yeast cells aren't terribly mobile (notice they have no feet, no arms, not even a tail), as they continue to reproduce, they clump.

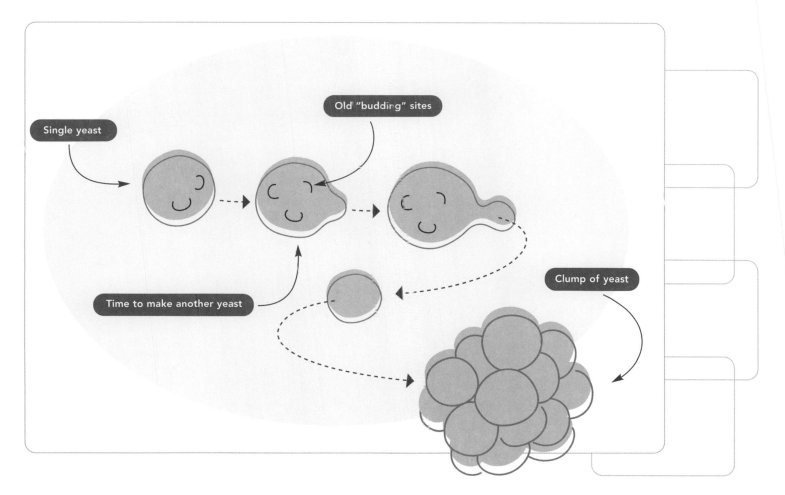

Eventually, the cells in the center of the clump start to die for lack of food and oxygen. The goal of "punching down" is to break up these clumps and help the yeast cells relocate to more spacious neighborhoods.

Instead of treating your dough like a punching bag, try this folding procedure:

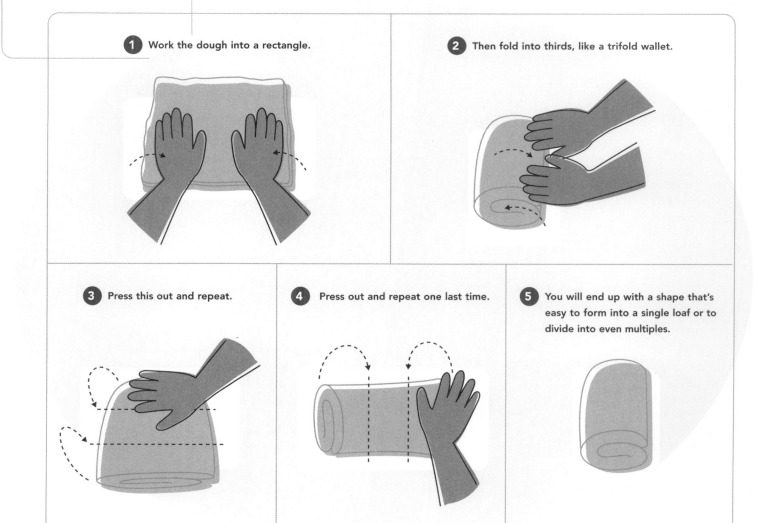

1. Work the dough into a rectangle.

2. Then fold into thirds, like a trifold wallet.

3. Press this out and repeat.

4. Press out and repeat one last time.

5. You will end up with a shape that's easy to form into a single loaf or to divide into even multiples.

Shaping. The goal of this step is to form the dough into the shape you want and to create a smooth skin on that shape, which is definitely the trickiest part of the program.

Forming a dough ball

1 Start by tucking the dough in on itself...

2 until a nice, smooth skin forms on top.

3 Roll the ball on the counter, between both hands, to tighten up the ball.

4 Then place in a well oiled bowl—seam side down—and swirl a few times to coat.

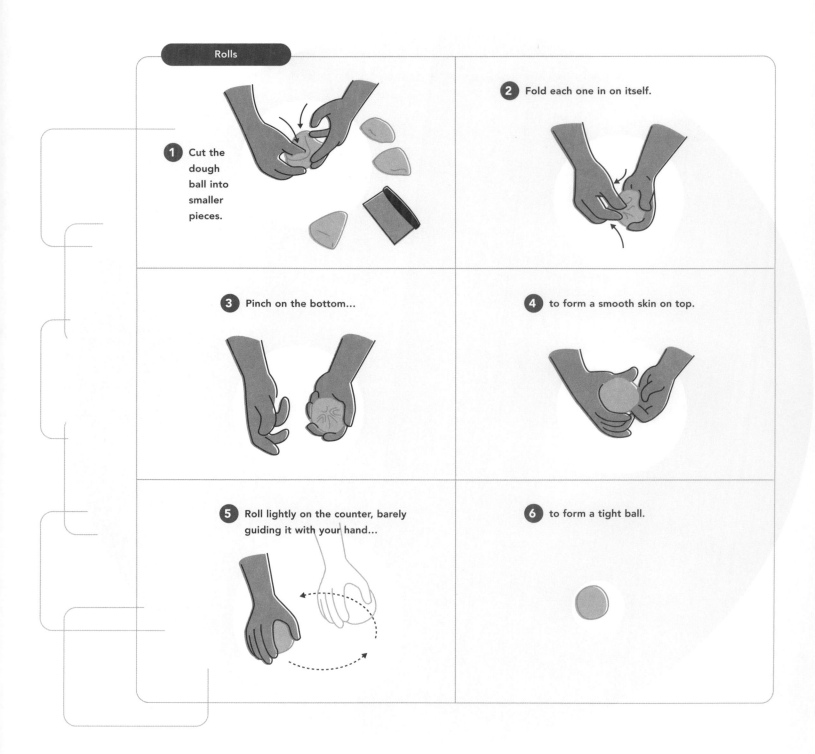

Rolls

1 Cut the dough ball into smaller pieces.

2 Fold each one in on itself.

3 Pinch on the bottom...

4 to form a smooth skin on top.

5 Roll lightly on the counter, barely guiding it with your hand...

6 to form a tight ball.

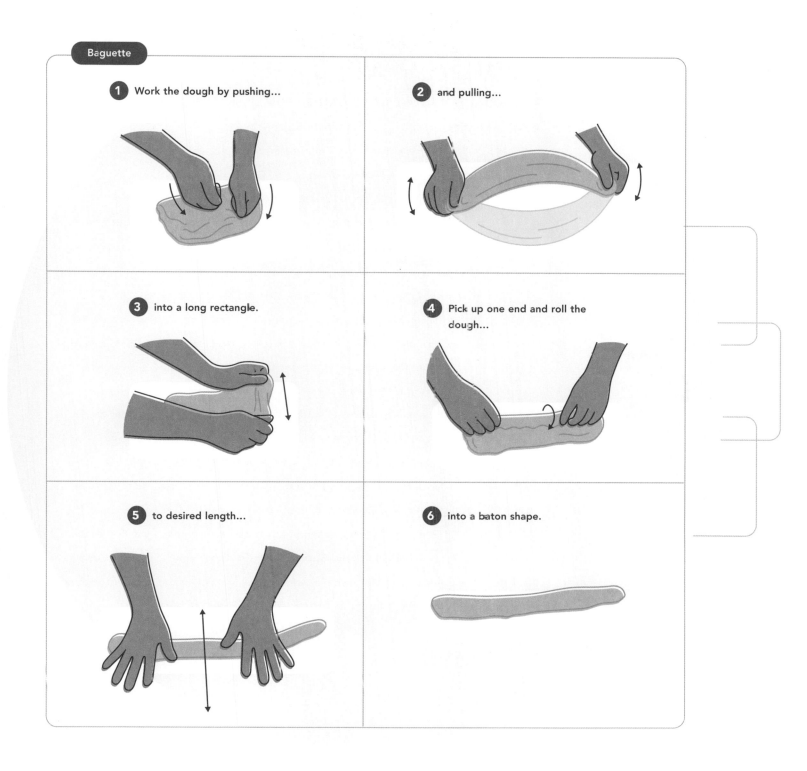

Baguette

1 Work the dough by pushing...

2 and pulling...

3 into a long rectangle.

4 Pick up one end and roll the dough...

5 to desired length...

6 into a baton shape.

Kneading Speeds

KNEADING DOUGH BY HAND CAN BE FUN, fill you with a sense of accomplishment, and help you build Popeye arms, but it's a heck of a lot of work. That doesn't mean I don't think it's worth doing, but I would hate for you not to make your own bread just because you didn't want to knead.

Prior to writing this book, I usually reached for a hand mixer whenever I needed to whip up egg whites or a cake batter. Although I still do reach for it from time to time, I have to admit I've become a fan of the electric stand mixer. There are several in our test kitchen but only two that can take care of business no matter what: the KitchenAid Professional 6 and the Viking Professional stand mixers. Despite some very nice design innovations on the part of the folks at Viking, the Pro 6 model is my favorite if for no other reason than that the Viking is just too freakin' loud. When testing the machine, I actually got into the habit of wearing my motorcycle earplugs. The noise is probably due to the fact that there is a hub on the back for a blender carafe. But I have to wonder, if I have a Viking blender carafe, wouldn't that probably mean that I have a Viking blender?

Anyway, back to speeds. The Pro 6 goes from Stir to 8, but I don't think we should concentrate on mixer speed numbers any more than we should be impressed by the fact that Nigel's amp goes to 11. When bringing batters together, I use Stir, or the lowest setting. When kneading, I use half-power (with the dough hook). I generally cream at 75-percent power; egg whites get whipped at half-power until frothy, then I drop the hammer and (affect Scottish accent here) give 'er all she's got.

The kneading of relatively stiff dough, such as pizza dough, is beyond the capability of most stand mixers, so if you're in the market, my advice is to buy the biggest one you can get.

Bench proofing. Depending on your recipe, this is either the second fermentation or the third. Some recipes may call for a second rise (again doubling the dough and then "punching" it down) before shaping or panning, but this final rise—known as the bench proof—takes place after the dough has been shaped.

You've just knocked a good bit of gas out of the dough, and this will give the yeast time to do a bit more work. If you skip this step, your finished pieces will be dense and...well, isn't that reason enough? I generally park the dough on my baking peel for this step, since moving it anywhere but to the oven when it's ready would misform the dough. And I usually cover the dough with a clean kitchen towel to keep it from drying out.

The amount of time the bench proof will take depends on the temperature of your kitchen, the composition of the dough, and the size of the pieces. A large loaf might take up to an hour, while smaller pieces might take only twenty minutes. You can also do this in the refrigerator over a longer period of time, just as you might the first fermentation. I once worked at a pizza place where the first fermentation took place at room temperature. The dough was then portioned, placed into shallow metal bowls, stacked, and refrigerated for at least eight hours prior to use. These days, I usually do it the other way around.

Allowing the bread to overproof is almost as bad—strike that, worse—than under-proofing it. Overproofing results in the yeast blowing up the bubbles too much—to the point that the dough is weak and can't support the rise. The loaves inflate like soufflés in the oven, only to collapse. It's ugly.

To test the dough development after bench proofing, give it another poke. Oddly enough, it won't bounce back (i.e. fill back in as shown on page 230). As for over-proofing at this stage, well, when I look at a product that's over-proofed, I think of a dead, bloated fish—yucky, but true. When you touch it, the outer membrane feels thin and flimsy—the firmness of the dough is all gone.

Slash and bake. Why slash a loaf? To give the dough somewhere to go. The skin on a shaped dough that's risen is more elastic than you might think. The inside wants to expand as the loaf heats, but more times than not, the outer shell doesn't want to budge. By slashing the bread by at least ¼ inch, you create fissures through which the hot bread can spread. Skip this step and your loaves will be very dense.

Slashing is traditionally done with a razor blade, but I find that a serrated bread knife does the job. Just make sure that you're not placing any downward pressure on the blade—literally drag it across the surface of the loaf.

Baking follows, and a lot happens in that oven: the yeast cells die; the bubbles expand; water turns to steam, further lifting the loaf; the starch gelatinizes and the proteins coagulate, providing structure; and the outside of the bread browns, thanks to the Maillard reaction.

When the bread comes out of the oven, this process continues for some time until the structure dries and sets. That's why you should never break into a loaf until it has had time to cool for at least half an hour. Rolls and pizza, possessing greater surface-to-mass ratios, don't need to cool as long.

Figurin' Yeast

AN ENVELOPE OF DRY YEAST—active dry or instant—measures up to 2½ teaspoons, give or take a few grains. Why manufacturers can't just go up to a tablespoon, I don't know. Truth be told, I don't like those little envelopes. I prefer buying yeast in jars or 1-pound foil bags. If you keep them tightly sealed and in the freezer, those little buggers will probably outlast you. (It's just like the cryo-sleep we always see in sci-fi movies.) An envelope of dry yeast also weighs 7 grams. That's about ¼ ounce.

Although many books and manuals present mathematical formulas for converting active dry amounts to instant, I just go 1:1. I know, this isn't technically exact, but I've never had a loaf go belly-up on me just because it had a little too much yeast in it. See what I mean about baking? It's no more an exact science than parenting. In fact, it's a heck of a lot like parenting. You have to get to know your kids in order to be any good to them.

The only conversion formula that I can think might come in handy is if you come up with a really old recipe that calls for compressed or "cake" yeast, which isn't really that common anymore. Although different bakers use different conversions, I would replace 1 ounce of compressed yeast with 1 envelope (7 grams or 2½ teaspoons) of instant yeast.

THE RIGHT AMOUNT OF FLOUR

♦ ♦ ♦

How do you know when you have the right amount of flour in the dough? By feel. This is why I always hold out some of the flour to work in on the board as I knead (and as needed). In the end, the dough should be smooth, soft, and just a wee bit tacky (as in sticky). When you stick a finger in it...well, you know the Pillsbury Doughboy? It should be just like his belly. When poked, the depression made by your finger should slowly fill in, leaving barely a dimple.

Depression slowly fills in on release

Well kneaded dough, indeed!

So here's my modification to the Straight Dough Method:

1. Scale or measure all ingredients.

2. Mix instant yeast and salt with all but 1 cup of the flour.

3. Add ingredients to work bowl, starting with the liquids and sugar (if called for), followed by the flour mixture.

4. Mix into a dough on low speed.

5. Cover the dough and allow it to rest for 30 minutes.

6. Turn out onto a board or other floured surface and knead until smooth and fully developed. Work in as much of the remaining cup of flour as is necessary to produce a smooth dough. Do not add more flour than called for in the recipe.

7. Cover and allow to rise until doubled.

8. Fold and press out the dough three times.

9. Rise again, or shape into final form.

10. Bench proof.

11. Slash and bake.

12. Cool.

THE SPONGE METHOD VARIATION

This is very similar to the Straight Dough Method, only it's split into two stages. First a very wet, loose dough is made using part or all of the liquid ingredients and the yeast, and only a third to half of the flour. This wet proto-dough—also known as the sponge—is left to ferment for some time (several days, if you like) before the final dose of flour—and the salt—is added. Then the dough is kneaded to the desired state. Once the sponge is fully developed, part of it can be reserved and maintained for use as a starter for future loaves.

The sponge method can be used with almost any yeast dough. The sponge rise time lengthens the time required for the recipe, but there is substantial payoff in increased flavor and refined texture. That's because the yeast had extra time to romp around in a wet, salt-free environment. Remember that salt restricts yeast fermentation via osmosis. And the added moisture makes it easier for the dough to reabsorb the gas—and flavors—produced by the yeast.

Here's how it works:

1. Scale or measure all ingredients.

2. Mix instant yeast with one half of the flour.

3. Add ingredients to work bowl, starting with the liquids and dissolvable solids—except the salt—followed by the flour/yeast mixture.

4. Mix into a dough, cover, and allow to rise until doubled in size.

5. Slowly add the remaining flour and all the salt.

SPONGE VS. STARTER: CONFUSED OR JUST CONFUSING?

◆   ◆   ◆

I've read about a zillion bread recipes and in doing so I've noticed that the terms "sponge" and "starter" are frequently used interchangeably. This is unfortunate, because they are not the same thing. A sponge is really just a step in the bread making process; that is, the mixture of a portion of the ingredients (including the yeast), which is then left to ferment for some time before the other ingredients are added. A starter is an active yeast culture that is maintained separately from any recipe; that is, it is an independent entity. A starter is designed to be used in place of commercial yeast and, though it often works more slowly than "instant" yeast, it brings a considerable amount of flavor to the party. I often use a starter in tandem with small doses of instant yeast so that I get both speed and flavor.

A starter can be cultivated from commercial yeast or from wild yeast; in the latter case, it's called a sourdough starter. In most cases, wild sourdoughs perform more vigorously than those formulated on commercial strains because wild yeasts are usually accompanied by wild bacteria that possess the enzymes necessary to break starch into the simple sugars that yeast prefer to munch. Wild yeasts can also thrive in highly acidic environments (starters become quite acidic over time) that would kill commercial strains.

WAYS TO KEEP A STARTER WARM

✦ ✦ ✦

A screened-in porch in summer would be perfect. If you don't have one, park the jar under a desk lamp, or wrap a heating pad set to low around the jar, secured with a rubber band or two.

The flour is added slowly here because by this point the sponge will have had time to soak up quite a bit of water. That means the dough will be less anxious to take on more flour. You may find that the prescribed amount of flour is just too much—and that's okay. I often find myself with a little flour left over. I just save it and slowly add it during the kneading. Remember that moist, pliable doughs rise better and in the end possess better texture than "hard" doughs that have been force-fed flour.

HERE ARE SOME GOOD STARTERS TO GET YOU STARTED:

All-Purpose Starter

This can be used in the construction of just about any yeast bread—strike that—any yeast bread.

In the carafe of a blender, combine

- ✦ 2 cups filtered or bottled water, heated to 110° to 120°F
  (just hot to the touch)

- ✦ 1 teaspoon sugar

- ✦ 1 teaspoon instant yeast

- ✦ 8 ounces bread flour

Blend for 10 seconds, or until smooth. Pour the mixture into a clean, quart-size glass jar. Barely screw on the lid—gases need to be able to escape. Skip this step and there'll be broken glass to pick up later.

Stash the jar in a warm (80° to 85°F would be nice) place for 24 hours. Use right away or feed it another cup of warm water mixed with a cup of flour and stash it in the refrigerator. Every time you use a cup of culture, feed the starter a cup of water and flour.

A Mason or Ball jar will serve nicely.

Wild Culture aka "Sourdough" Starter

Wild yeast can be cultivated from organic flour (rye flour is well known for supporting large yeast populations, but any organic will do), organically grown grapes, or thin air. I've had success with combining one and three.

In a bowl, crock, or other vessel, mix together

- 1 cup organic bread flour

- 2 cups warm filtered or bottled water

- 1 teaspoon sugar

Place the bowl—uncovered—in an open-air, bug-restricted spot where, ideally, the temperature remains between 75° and 95°F. Let the mixture sit for 3 days and wait for the fun to happen.

Within that time, trillions of yeast cells will come to roost, along with some of their bacterial buddies. You'll know you have goo-guests when the mixture starts to slowly bubble (that's CO_2) and get stringy when stirred. If all goes well, in 8 to 10 days you'll have a unique sourdough starter capable of producing bread that tastes specifically of wherever you are. If you add a cup of water and a cup of flour every time you remove a cup of starter, you should be able to keep it going indefinitely. I have a friend who says she knows a little boulangerie in Paris that claims to have kept their starter going since the time of Napoleon I. I suspect this is only clever marketing, but technically, it's doable.

Note: Wild cultures can be taken over by foul-tasting or even harmful microbes. Watch for odd color changes, especially in the hooch. If the liquid turns darker, or becomes reddish or orange, dump the whole batch and start over. (Yet another reason to use commercial, yeast-based liquid culture, if you ask me.)

The most bread-friendly bacteria prefer cool weather; hot weather brings out less-tasty bugs.

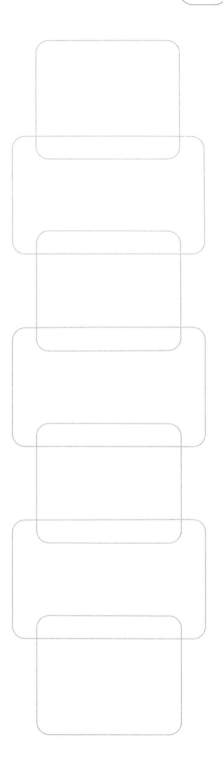

Got Milk? Then you have alcohol, too.

Starters both wild and tame can be built on milk as well as water. Milk-based cultures produce a more sour, yogurty flavor because naturally occurring bacteria will convert some of the milk sugar, or lactose, into lactic acid. Some such starters take on a winelike aroma, which is just fine if you like that kind of thing. It's important to note that the amount of alcohol we're talking about here is minuscule, but crucial.

The alcohol produced by baker's yeast is important because alcohol evaporates at a lower temperature than water—which means that a dough containing some alcohol will spring faster in the oven than a dough that doesn't contain alcohol. Don't worry—an insignificant amount of alcohol remains behind in the finished bread.

Milk Starter

This is very good for Everyday Bread (page 252).

In a microwave-safe vessel, combine

✦ 1 cup warm water

✦ 1 cup skim milk

Heat the mixture in a microwave oven until it reaches 110° to 120°F.

In the work bowl of a food processor, combine

✦ 1 package (2½ teaspoons) of yeast

✦ 2 cups bread flour

Pulse to sift them together. Add the warm liquids to the flour mixture and pulse just to combine. Move the mixture to a 2-quart glass or polycarbonate (Lexan) container. Cover loosely (gas needs to be able to escape) and hold at 85°F for 24 to 48 hours, stirring every 6 hours or so.

After two days, the starter should smell nice and sour and be ready to use. Always stir well before removing a portion for use and always feed the starter with a cup of warm skim milk and a cup of flour for every cup of starter used. Stir well, cover, and stash in a warm place until bubbly, then return it to the refrigerator.

The Last Word: Dry Yeast, Sponge, Starter

A bread constructed via the Straight Dough Method using dry yeast only will be just fine. The flavor of the flour and other ingredients will come through. A good example of this method is a standard baguette—it's bread, and it tastes good.

A bread constructed via the sponge variation will have considerably more flavor due to the extra time that the yeast cells have had to eat, live, reproduce, die, and so on. If you want more flavor in your bread, have at least an hour of extra time to spare, and are not willing to keep a starter on hand, this is definitely the best way to go.

A bread constructed via the Straight Dough Method using starter alone will have a deep, tangy, complex flavor and a slightly chewier texture. If the necessary portion of starter is removed from the chill chest and allowed to come to room temperature prior to mixing the rest of the dough, no significant extra time will be required. In my experience, starters alone can be inconsistent when it comes to gas production, so I usually add at least a quarter teaspoon of instant yeast to the flour when I assemble the dough.

THE CARE AND FEEDING OF STARTERS

◆ ◆ ◆

Shake or stir your starter often. Yeast cells aren't very mobile, so they form clumps as they reproduce. Agitation helps to redistribute the critters through the culture, where they can breathe and eat and, yes, make more yeast.

If you use your starter once every two weeks, it should last forever as long as you feed it. If your starter goes unused for a month, bring it to room temperature, throw away all but a cup, and start over by feeding it a cup of warm water and a cup of flour. And once you have a culture going, never change its food. If you start it on water and flour, don't switch to milk and flour.

Rolling Pins, Stones, and Peels, Oh My!

OF ALL THE SPECIALIZED TOOLS designed for bakers, there are only three I wouldn't want to try living without: a French rolling pin, a baking stone or two, and a baker's peel.

The rolling pin is good for a zillion and one tasks, including cracking nuts, crushing graham crackers, coldcocking burglars, and yes, even rolling dough.

A baking stone has two jobs: to soak up heat and divvy it up to whatever happens to be sitting on it. Placing a good one in the bottom of your oven will go a long way toward improving your loaves, be they free-form or panned. And since the stone has a lot of mass, it helps even out the ups and downs of your oven, which is why I never take mine out.

Although you can buy a "pizza stone" at just about any decent cook's store, I prefer using big, unglazed quarry tiles. If such things can't be gotten on the cheap from a builder's supply in your area, see if there's a pottery supply store around town. They'll probably have ceramic shelves for large kilns, and those make dandy stones. If you have some extra cash lyin' around, you can consider going to a counter cutter and having a piece of soapstone cut to fit your oven.

How big a stone do you need? Before going on a Neolithic hunt, take a close look at the bottom of your oven—and grab a tape measure while you're at it. Is the floor of your oven solid? Are there little air vents around the perimeter? Or does a long Calrod unit snake its way across the floor? If the floor is completely solid with no vents of any type, odds are you have an electric oven with a covered element and you're in luck. You can get a stone (or stones) that completely covers the bottom of the oven, and you can place it flush to the floor. Got vents? Then you have a gas oven, and you'll need to make sure that the stone you place on the floor doesn't cover any of those holes—not even by a little. They're important. If you have an exposed Calrod, you'll have to set your stone on a rack in the bottom (A) position. The stone can be a large as you like, as long as it doesn't interrupt the operation of the grate.

As for the thickness of the stone, mass is good—mass works. Try to avoid oven stones that are less than 1 inch thick. Are two better than one? You bet they are. By placing one stone on the floor of the oven and another on the top rack, you can imitate the heat-radiating characteristic of a wood-burning oven. You can't produce the same heat of course, but that's the way it goes.

Steam on stones. Adding steam to the oven during the first few minutes of baking can convert an okay crust to a truly great crust. Before I got my stones, I just tossed a cup of water onto the floor of the oven. But with the stone on the floor and the bread on the stone, there was no longer any place to throw that water. If you have two stones, just make sure that the top one is positioned so that you can get the water onto it. If the oven's really hot, you won't have to worry about water rolling off the edge of the stone—it'll never get that far. If you don't have a second stone, or you have an oven with an exposed Calrod, place a shallow pan (like a jelly roll pan) on a rack in oven position A. Your stone should be placed on a rack that's in position B or C (it should be in the center). As soon as the bread hits the stone, pull out the rack on the bottom and pour ½ cup water into the pan.

There is no better way of moving baked goods into or out of the oven than a baker's peel. Whenever possible, I do my final loaf shaping right on the peel, so that when the pieces have been proofed, all I have to do is slash and slide. The key to the slide is cornmeal (or grits or polenta). Strewn lightly between peel and bread, the hard bits act like ball bearings, allowing the bread to slide on and off the wooden platform easily. Meal won't be softened by long exposure to wet doughs, though you may have to use a lit-tle more to ensure the dough doesn't stick to the board.

Moving foods from peel to stone and back again without mishap will require some practice. I suggest you buy a few frozen pizzas and get to work.

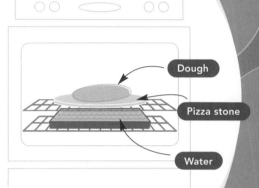

Dough

Pizza stone

Water

Dangerous? Maybe, but I don't see how. I'll ask my lawyer.

"Peel" comes from the Middle English *pele*, from the Middle French, from the Latin *pala*, meaning spade.

Hardware:

Digital scale

Dry measuring cups

Wet measuring cups

Measuring spoons

Stand mixer with paddle attachment and dough hook

Large stainless steel or glass bowl

Plastic wrap

Pizza stone or 12 x 12-inch smooth ceramic tile

Chef's knife or bench scraper

2 clean kitchen towels

Baker's peel

Notes:

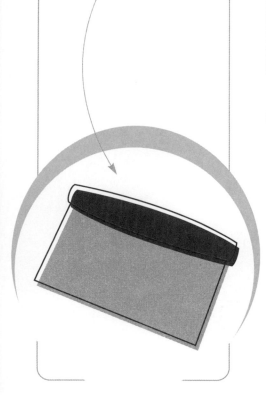

Pizza Dough

This recipe never lets me down. I never have to add additional flour or water and the overnight rise always delivers the flavor and texture I'm after. Although I developed this specifically for pizza and other flatbread applications, it makes great breadsticks and rolls—even small, torpedo-shaped loaves. It can even be coaxed into crackers if properly docked. If you only memorize one application from this book, make it this one.

THE LIQUID

INGREDIENT	Weight		Volume	Count	Prep
Water	283 g	10 oz	1¼ cups		warm

THE DRY GOODS

INGREDIENT	Weight		Volume	Count	Prep
Chewable children's vitamin C	25 mg			1	crushed
Salt	11 g	½ oz	1 tablespoon		
Sugar	5 g	⅙ oz	1 teaspoon		
All-purpose flour	454 g	1 lb	4 cups		
Instant yeast	7 g	¼ oz	2½ teaspoons	1 envelope	

THE EXTRAS

Olive oil for the bowl ✦ Cornmeal for dusting the peel ✦ Toppings of your choice

✦   ✦   ✦   ✦   ✦

✦ Dissolve the vitamin C in the warm water, then add to the work bowl of an electric stand mixer fitted with a dough hook. Add the remaining ingredients (reserving some of the flour) and mix according to the **STRAIGHT DOUGH METHOD.** Mix for 2 minutes on low, or until the dough pulls away from the sides of the bowl, adding more flour by the tablespoon as needed.

✦ Rest the dough for 15 minutes, then knead at 35- to 40-percent power for 5 minutes, or until the dough is well developed.

✦ Remove the dough from the mixer bowl and knead by hand for about 30 seconds, then work the dough into a ball. Place the ball in a large metal bowl with a little olive oil. Toss the ball around to coat. Cover the bowl with a clean kitchen towel and set aside for one hour or until the dough ball nearly doubles in size.

✦ Fold down the dough, patting it into a disk, and place it back in the bowl. Cover with plastic wrap and park it in the fridge overnight.

✦ The next day, remove the dough and cut it into four equal portions. Shape these as you did the mother blob, folding the dough in on itself. If you plan to have pizza that day, leave however many orbs you like on the counter, cover them with a clean kitchen towel and leave for an hour to bench proof. The others should be wrapped in plastic wrap, or stored in zip-top bags, in the refrigerator for up to a week. If you want frozen pizza, you're better off rolling out and par-baking the crust, then freezing. Thaw the frozen crusts before you finish them.

✦ While the dough is proofing, preheat your oven to its highest possible temperature (or light up the grill).

✦ Prepare the dough:

✦ Dust the peel with cornmeal and place your dough on top—and top it. I keep most of my pizzas very simple: olive oil, a little cheese, a few toppings like herbs and olives. I don't use tomato sauce very often.

✦ Slide the pizza onto the pizza stone in the oven. These pizzas don't take more than 4 to 5 minutes to become bubbly and golden brown. Allow to rest for 3 minutes before slicing.

Yield: This recipe makes 726 grams of dough, or roughly 25 ounces, which makes four 6.25-ounce pizzas (a great 1- to 2-person size).

The higher and faster the dough spins, the more it will increase in diameter. If you've never done this, don't get too crazy—just try getting the dough a foot or so off your hands. Each time it lands (preferably on your hands rather than the floor) move your hands apart to stretch it a little more. You should be able to work this amount of dough out to the size of a dinner plate and that's more than enough.

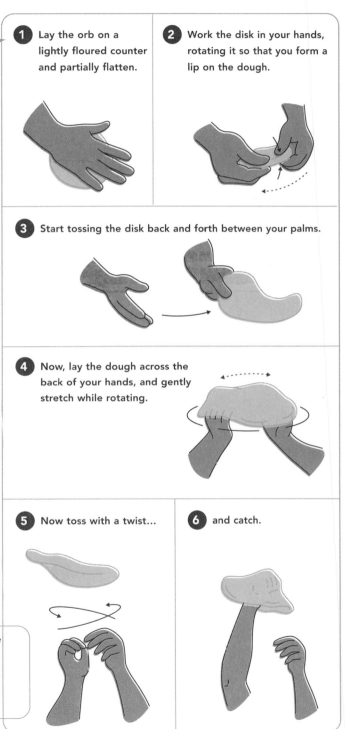

1 Lay the orb on a lightly floured counter and partially flatten.

2 Work the disk in your hands, rotating it so that you form a lip on the dough.

3 Start tossing the disk back and forth between your palms.

4 Now, lay the dough across the back of your hands, and gently stretch while rotating.

5 Now toss with a twist...

6 and catch.

From Pizza to Brioche

Pizza dough is strong and potentially as elastic as the waist-band on a new pair of BVDs. Brioche, on the other hand, is as soft and tender as lingerie. These two classic breads are as different as day and night on the plate, but when you look under the hood they are surprisingly alike. All it takes is a little science to move from one to the other.

First let's review some characteristics:

PIZZA	BRIOCHE
Lean, no fat	Very rich and buttery
Interior color: very light	Yellow
Not at all sweet	Mildly sweet
Chewy, crusty	Tender but not necessarily soft; can hold its form even when cooked into a bread pudding
Exterior color: lightly browned	Deep mahogany
Texture open, coarse	A little more refined

Here's the ingredient list again for my pizza dough:

INGREDIENT	AMOUNT
Flour	1 pound
Salt	1 tablespoon
Yeast	1 pack
Sugar	1 teaspoon
Water	10 ounces
Total moisture	10 ounces

Okay, let's say for a moment that we're going to stick with a pound of flour and a tablespoon of salt. Where might we go from here? Well, I'm not going to lie to you. I had a few missteps along the way, but after three batches, here's what I came up with.

Now I like my pizza dough, and in order to make it pliable but workable it requires about 10 ounces of total moisture. But with brioche I know I need to get it from somewhere besides the tap. So step one: Dump the water, keeping in mind that in the end I'm going to need to get that 10 ounces back.

Notes:

So now I'm looking at:

INGREDIENT	AMOUNT
Flour	1 pound
Salt	1 tablespoon
Yeast	1 pack
Sugar	1 teaspoon
Water	0
Total moisture needed	10 ounces

Having looked over a few traditional brioche recipes, I know that they often call for half as much butter as flour.

So I add 8 ounces to the formula:

INGREDIENT	AMOUNT
Flour	1 pound
Salt	1 tablespoon
Yeast	1 pack
Sugar	1 teaspoon
Butter	8 ounces
Total moisture	1.2 ounces

Notice that the moisture count now stands at 1.2 ounces. That's because I'm figuring the butter at 15 percent water. Of course, to add richness, color, and some structure to bolster all that fat, we'll need eggs. Four of them.

Now I have:

INGREDIENT	AMOUNT
Flour	1 pound
Salt	1 tablespoon
Yeast	1 pack
Sugar	1 teaspoon
Butter	8 ounces
Eggs	4 large
Total moisture	8.2 ounces

Notice the moisture is up to 8.2 ounces. That's because one large egg weighs about 1.8 ounces. Four of them would weigh 7.2 ounces and I figure about 7 of ounces of that weight is water. We still need a little more liquid, preferably one that would bring some browning to the crust. I'm figuring 1.25 ounces of milk should do the trick.

That gets us up to:

INGREDIENT	AMOUNT
Flour	1 pound
Salt	1 tablespoon (.4 ounces)
Yeast	1 pack (.250 ounce)
Sugar	3 tablespoons (1.5 ounces)
Butter	8 ounces
Milk	1.25 ounces
Total moisture	9.45 ounces

That's close enough to 10 ounces for me. Whether or not this added amount will be necessary will depend on a variety of factors, So how do you know if you need it? Keep reading.

Notes:

Actually, 1.75 ounces, but I'm rounding up.

Notes:

Now, brioche is considerably sweeter than pizza dough, so I'm going to multiply the teaspoon of sugar that was in the pizza, which was just enough to tempt the yeast, by a factor of 9, bringing us up to a whopping 3 tablespoons. You'll no doubt remember from the ingredients section that sugar is a tenderizer, but that's okay, because the egg proteins are builders and will make up for whatever the sugar tears down. Last but not least, brioche dough is heavy and dense because of all the added fat, so I'll double the yeast for more lift.

So here is the full conversion of the formula...sorta Jekyll to Hyde:

INGREDIENT	PIZZA	BRIOCHE
Flour	1 pound	1 pound
Salt	1 tablespoon	1 tablespoon
Yeast	1 pack	2 packs
Sugar	1 teaspoon	3 tablespoons
Butter	0	8 ounces
Eggs	0	4 large
Milk	0	1.25 ounces
Water	10 ounces	0
Total moisture	10 ounces	9.25 ounces

So on to brioche production, which just so happens to be a fine example of a Two-Stage Dough (but with a twist). Many yeast breads are assembled in two parts. First, a wet, doughy mass is mixed together from part of the formula's flour, most or all of its moisture (including eggs), and all the yeast. Sometimes this proto-dough is exclusive to a particular application, like this brioche. Other times it's a generic starter (wild or tame), which can be kept on hand for use in a wide range of goods. This procedure can serve a few different purposes. In the case of a sourdough bread featuring a wild starter, the issue is flavor. In the case of this brioche it's more about giving the yeast a running start at a challenging lift. By allowing fermentation to begin before the dough is developed, a bread that might not rise to the occasion, can.

Brioche procedure

✦ In the work bowl of an electric stand mixer, combine

✦ half the flour

✦ all the eggs mixed with all the milk

✦ all the yeast

✦ Mix on low speed using the paddle. When the mixture appears homogenized, turn off the machine, cover the work bowl with plastic wrap (no need to remove it from the mixer), and walk away for an hour.

✦ Mix the rest of the flour with the salt and sugar, add it to the dough, and mix again. When the dough comes together, change to the bread hook and knead the dough on 30 percent power until smooth and glossy. You'll probably have to scrape down the hook at least once during this time. When the dough is smooth, plastic, and elastic (kinda like a baby's bottom), shut down the machine and let the dough rest.

This gives you time to prep the butter.

Croissant dough, puff pastry, and brioche all have one thing in common: They contain butter that's integrated while solid. Both croissant and puff dough are laminates. That is, the dough is layered with many, many layers of butter. Brioche is easier to make because the butter is simply beaten into the dough. Problem is, to do it right, we're talking about beating solid butter into a dough that's already fully developed. To do that, we need butter that's plastic but not soft on the verge of melting.

✦ So, peel two sticks of butter, stack them on top of each other and beat them together (gently) with a wooden rolling pin. Sprinkle flour on the counter and roll the butter blob in it. (If you don't, the butter will stick to the rolling pin).

✦ Now quickly beat the butter into a rectangle about ¼ inch thick. Using a bench scraper or metal spatula, fold the butter over on itself like a tri-fold wallet and beat it out again. Fold and pound one more time, then cut the butter into strips.

Notes:

Brioche can be assembled by hand, but, boy, is it tough.

In this case the first fermentation included the autolyse. This stage is just to rest the gluten so that the butter can be worked in more easily.

Notes:

Now, if there's a trick to brioche, it's how you get the butter into the dough without it melting. If you simply dump it into the mixer bowl a piece at a time while the hook is in place, the dough will just smash the butter against the side of the bowl. That will only serve to lube the dough ball. Remember, in order to actually work the dough, the hook, paddle, whatever, has to work with the bowl…there has to be friction. If the dough ball is covered in butter, friction won't be much of an option.

My KitchenAid Pro 6 has enough torque to pop a finger out of its socket (I'm guessing). So when I make brioche, I rest the developed dough and then I put the paddle back in and turn it to 30 percent power. The motor moans, but turn it does, and the paddle cuts into the dough enough so that I can the drop the butter in a piece at a time and have it all worked in within about two minutes. Once half the butter is in, the dough loosens up a little, but until then it's a real struggle. If I had a lesser mixer, this would not be possible. Had I a lesser mixer, I'd be doing one of two things:

✦ Kneading the butter in by hand. This is best done by vampires. Not only are they strong (so I hear), but they are darned cold, and that's the big issue when you're trying not to melt butter that's already in a plastic state.

✦ I'm going to assume that you're not a vampire and advise you to do the following:

1 Take the rested dough out of the mixer and roll it out as thin as you can with a rolling pin.

2 Once you've got it rolled out, place the butter on it like this.

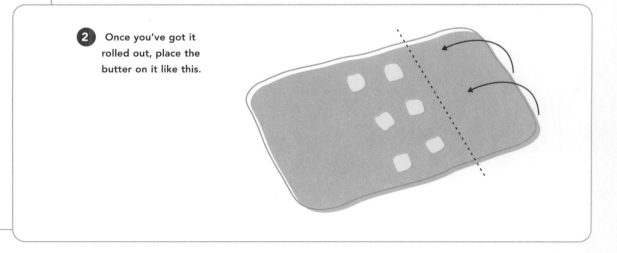

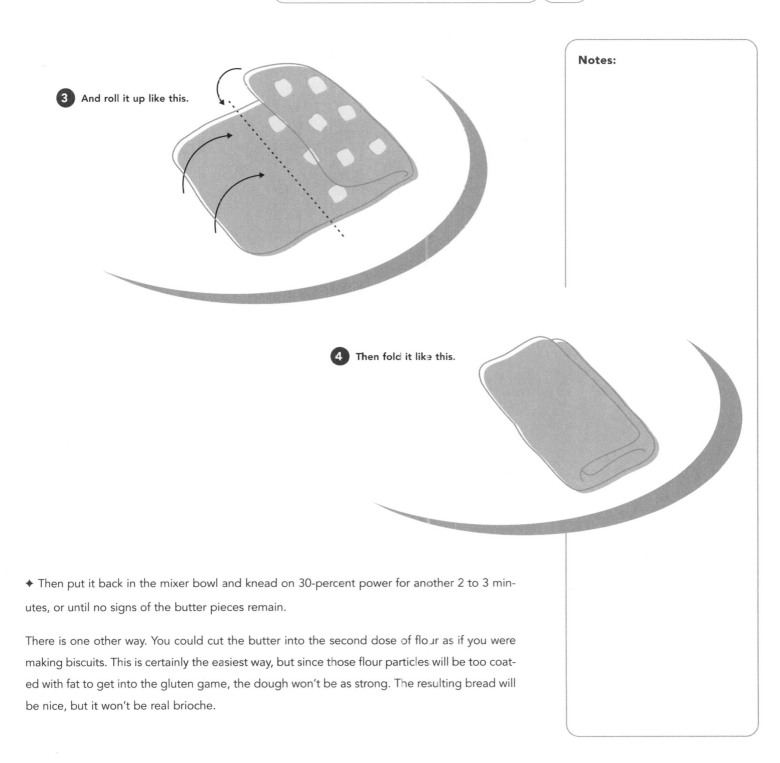

Notes:

3 And roll it up like this.

4 Then fold it like this.

✦ Then put it back in the mixer bowl and knead on 30-percent power for another 2 to 3 minutes, or until no signs of the butter pieces remain.

There is one other way. You could cut the butter into the second dose of flour as if you were making biscuits. This is certainly the easiest way, but since those flour particles will be too coated with fat to get into the gluten game, the dough won't be as strong. The resulting bread will be nice, but it won't be real brioche.

Notes:

✦ Once the dough is constructed, roll it into a ball, just as you would pizza dough, and stash it in the chill chest overnight.

✦ In the morning, prep either two 8½ x 4½-inch loaf pans or one 13 x 4-inch Pullman loaf pan (see pages 180–183), which is just like a loaf pan only longer.

✦ Cut the dough into 16 equal pieces and quickly roll them into balls. Then just pile them into the pan. Cover and allow to rise at room temperature for 3 hours.

✦ When the time is up, crank the oven to 375°F. When the oven reaches its goal, place the loaf or loaves in the middle of the oven (rack in position C) and bake for 40 minutes or until the internal temperature reaches 190°F.

✦ Cool 15 minutes in the pan then turn out and cool on a rack.

✦ Slice and enjoy with nothing at all—okay, maybe a little jam. Believe me, you won't need any butter. Also makes great French toast, though for that you'll need stale bread, and since it contains so much sugar (which is hygroscopic) staleness won't happen very quickly. Leftover brioche also makes great bread pudding. While fresh it can be served with ice cream just like cake.

So Why Go Through All This?

WHY TAKE ANY JOURNEY? Why did the chicken cross the road? Because in moving from one place to another, we learn something about both. Maybe you're not the kind of person who needs to know how things work, but if you have your eyeballs on this page, I'm betting that you are. And people like us tinker—that's how we find things out. For me, following the trail from pizza to brioche just shows the lay of the land. Seeing how everything is connected rather than just following directions makes me a better cook.

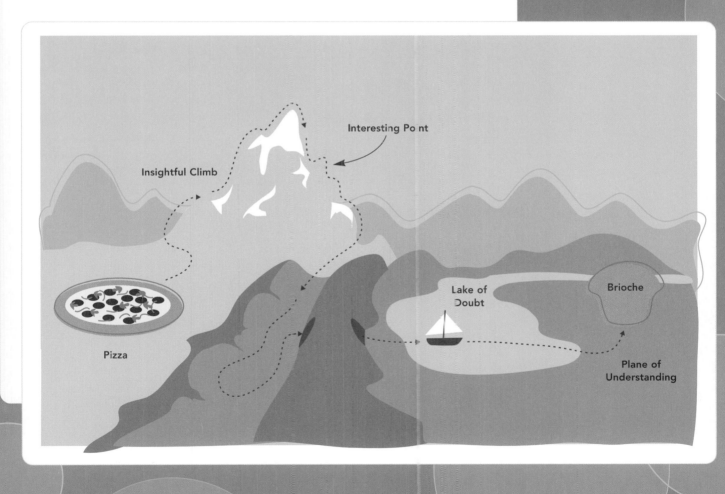

Potato Rolls

No one said that all the starch in a bread has to come from wheat flour. Because the starch composition of potatoes is different than that of wheat, these rolls have a unique texture—and yes, they taste ever so slightly of potatoes.

THE LIQUID

INGREDIENT	Weight		Volume	Count	Prep
Whole milk			1 cup		
Unsalted butter	43 g	1½ oz	3 tablespoons		melted

THE DRY GOODS

INGREDIENT	Weight		Volume	Count	Prep
All-purpose flour	373 g	13⅛ oz	2¾ cups		
Potato flakes	20 g	⅝ oz	⅓ cup potato flakes		
Kosher salt	5 g	¼ oz	2 teaspoons		
Sugar	42 g	1½ oz	3 tablespoons		
Instant yeast	7 g	¼ oz	2½ teaspoons	1 envelope	

THE EXTRAS

Vegetable oil for the work surface ✦ Melted butter to brush the rolls ✦ Parchment for the pan

Hardware:

Digital scale

Dry measuring cups

Wet measuring cups

Measuring spoons

Small saucepan

Whisk

Stand mixer with dough hook

Large bowl

Clean kitchen towel

One 9 x 13-inch sheet pan or two 9-inch round cake pans

Pastry brush

Notes:

I'm a big fan of potato flakes. I use them to thicken soups and stews.

✦ ✦ ✦ ✦ ✦

✦ Combine the milk and butter in a small saucepan and heat until the mixture reaches 110° to 120°F.

✦ Mix together the flour and the potato flakes, then assemble the dough via the **STRAIGHT DOUGH METHOD** in the work bowl of an electric stand mixer.

✦ Knead the dough until it's fully developed, about 5 to 10 minutes if kneading in the mixer, 7 to 10 minutes by hand. The dough will become shiny and elastic.

✦ Allow the dough to rise in a covered bowl for 1 to 2 hours, until it is doubled in bulk.

✦ Turn the dough out onto an oiled work surface and divide it into sixteen 1½- to 2-ounce pieces. Form into balls or knots and place them, about 1 inch apart, in a 9 x 13-inch pan or two 9-inch round cake pans.

✦ Place a clean kitchen towel over the rolls and allow them to rise, covered, for 1 to 2 hours.

✦ When the bread is just about ready to go, place an oven rack in position C and preheat the oven to 350°F.

✦ Bake until the temperature reads 190° to 195°F on an instant-read thermometer, or until the rolls are a light golden brown, about 20 to 25 minutes. Brush the rolls with more melted butter after removing them from the oven.

✦ Serve warm.

Yield: 16 dinner rolls

WHAT'S YOUR KNEADING STANCE?

✦ ✦ ✦

I prefer to knead right on my Corian table top. Wood is fine, too. There is no right or wrong surface for kneading, although I suppose gravel would be tough. One thing I do notice, though, is that Americans don't like to get their hands sticky——so they over-flour their surfaces. That often ends up producing hard loaves.

When kneading, consider changing your stance.

Counter height:
not much leverage.

Kitchen table:
allows for more downward pressure.

Hardware:

Digital scale

Dry measuring cups

Wet measuring cups

Measuring spoons

Electric stand mixer with
dough hook

Large bowl

Clean kitchen towel

Baking sheet

Instant-read thermometer

Cooling rack

Notes:

Everyday Bread

I bake this bread every week or so, and it never lets me down. Because I use a starter (which I've had around for quite a while now), this bread goes together relatively quickly. The taste is tangy but not too assertive…nothing pushy.

THE LIQUID

INGREDIENT	Weight		Volume	Count	Prep
Water	397 g	14 oz	1³⁄₄ cups		filtered and warmed to 110° to 115°F

THE DRY GOODS

INGREDIENT	Weight		Volume	Count	Prep
Sugar	5 g	¹⁄₆ oz	1 teaspoon		
Starter	454 g	1 lb			
Flour	680 g	24 oz	5 cups −2 tablespoons		
Kosher salt	3 g	¹⁄₈ oz	1 teaspoon		

THE EXTRAS

INGREDIENT	Weight		Volume	Count	Prep
Water	43 g	1¹⁄₂ oz	3 tablespoons		
Cornstarch	5 g	¹⁄₆ oz	1 teaspoon		

That's right—you need to weigh it.

No chlorine, please.

You can use any of the following combinations of flour. Each will produce very different loaves.

- 24 ounces all-purpose (light and white but nicely tangy)
- 12 ounces all-purpose + 12 ounces bread flour (a bit more chew)
- 12 ounces all-purpose + 6 ounces bread + 6 ounces wheat flour (heartier still)

✦ ✦ ✦ ✦ ✦

✦ Combine all ingredients via the **SPONGE METHOD VARIATION.**

✦ For a coarser, more rustic bread, fold the dough down and proceed as follows. For a finer texture, allow the dough to have a second rise overnight in the refrigerator, and then proceed to the next step.

✦ Roll the dough between your hands on a counter or cutting board until a tight, smooth ball is formed. Place the dough ball on the baking sheet.

✦ Cover with a clean towel and bench proof until doubled in size. Place an oven rack in position B and preheat the oven to 400°F.

✦ Meanwhile, pour 3 tablespoons of water into a small saucepan, and add 1 teaspoon of cornstarch. Heat just until the cornstarch dissolves, but not to the point that it thickens the water. Remove from the heat.

✦ When the bread is risen again, paint on the cornstarch slurry with a brush, then slash the top in a box pattern, cutting only about ½ inch into the loaf.

✦ Bake for 40 minutes—the interior should be 190° to 195°F in the center, and the outer crust will be quite brown.

✦ Cool on a rack. You'll notice that the outer shell makes a Snap, Crackle & Pop sound as the crust cracks from the contraction of the starch. My daughter likes to say that the loaf is "talking."

Notes:

450° for a coarser crust.

Pillow Bread

This is a rustic bread with a light, delicate crumb.

THE LIQUID

INGREDIENT	Weight		Volume	Count	Prep
Milk	340 g	12 oz	1½ cups		
Water		13 oz	1⅔ cups		divided
Honey	85 g	3 oz	6 tablespoons		
Unsalted butter	57 g	2 oz	4 tablespoons		melted

THE DRY GOODS

INGREDIENT	Weight		Volume	Count	Prep
Cornmeal	75 g	2⅝ oz	½ cup		
Salt	19 g	⅔ oz	4 teaspoons		
Bread flour	875 g	1 lb 15 oz	6½ cups		
Instant yeast	15 g	½ oz	1½ tablespoons		

THE EXTRAS

Baker's Joy or AB's Kustom Kitchen Lube for the pan

Hardware:

Digital scale

Dry measuring cups

Wet measuring cups

Measuring spoons

Small saucepan

Whisk

Stand mixer with dough hook

Large bowl

Half sheet or jelly roll pan

Clean kitchen towel

Instant-read thermometer

Cooling rack

Notes:

Notes:

✦ ✦ ✦ ✦ ✦

✦ Bring 1 cup of water to a boil in a small saucepan. Slowly whisk in the cornmeal and cook, whisking constantly, until the mixture thickens, 30 seconds to 1 minute.

✦ Allow the cornmeal to cool thoroughly and add it to the flour in a large bowl.

✦ Combine the milk, remaining water, honey, and butter in a small saucepan and heat until the mixture reaches 105° to 115°F.

✦ Assemble the dough via the **STRAIGHT DOUGH METHOD.**

✦ Knead for 8 to 10 minutes, until the dough has fully developed.

✦ Form the dough into a loaf on a prepared (see pages 180–183) half sheet pan, cover with a clean kitchen towel, and allow to rise for 1½ hours.

✦ When the bread is just about ready to go, place an oven rack in position C and preheat the oven to 350°F.

✦ Bake until the temperature reads 190° to 195°F on an instant-read thermometer, or until the loaf is slightly golden on top, about 20 to 25 minutes.

✦ Remove immediately from the pan to a cooling rack. Allow the loaf to cool completely before slicing.

Yield: One round rustic loaf

Hardware:

Digital scale

Dry measuring cups

Wet measuring cups

Measuring spoons

Stand mixer with dough hook

Large bowl

Plastic wrap

9 x 5-inch loaf pan

Serrated bread knife

Instant-read thermometer

Cooling rack

Notes:

Multigrain Loaf Bread

This is the perfect place to apply your liquid culture.

THE SPONGE

INGREDIENT	Weight		Volume	Count	Prep
Bread flour	135 g	4$\frac{3}{4}$ oz	1 cup		
Whole wheat flour	72 g	2$\frac{1}{2}$ oz	$\frac{1}{2}$ cup		
7-grain cereal	52 g	1$\frac{3}{4}$ oz	$\frac{1}{2}$ cup		broken up fine
Starter	227 g	8 oz	1 cup		
Water	113 g	4 oz	$\frac{1}{2}$ cup		
Buttermilk	113 g	4 oz	$\frac{1}{2}$ cup		full-fat, if possible

THE FINAL DOUGH

INGREDIENT	Weight		Volume	Count	Prep
Olive oil	13 g	$\frac{1}{2}$ oz	1 tablespoon		
Honey	43 g	1$\frac{1}{2}$ oz	2 tablespoons		
Kosher salt	7 g	$\frac{1}{4}$ oz	1 tablespoon		
Bread flour			pinch		
Whole wheat flour	113 g	4 oz	$\frac{3}{4}$ cup		

THE EXTRAS

2 additional ounces bread flour to be used only if finished dough is sticky

Baker's Joy or vegetable-oil spritz

✦   ✦   ✦   ✦   ✦

✦ In the work bowl of a stand mixer, stir together all of the ingredients for the Sponge. Cover loosely and allow to double in volume. This may only take a few hours at room temperature, or overnight if you put it in the fridge. I prefer the overnight method because I like to do the actual baking in the morning while the house is still cool.

✦ To make the final dough, place the work bowl containing the sponge onto the mixing stand and fit the mixer with the dough hook. Turn the mixer on to Stir (the lowest speed) and add the olive oil, honey, and kosher salt.

✦ In a separate container, combine the bread and whole wheat flours.

✦ Add the flour mixture to the sponge in doses until the dough looks...well, kind of like a shaggy mess. You don't want the dough to be wet, but if it's too dry, the bread won't be sandwich-soft, either. So if you find yourself holding back a little on the flour, that's okay.

✦ Boost the mixer to low and knead for 5 minutes, or until the dough is smooth and springy. Turn off the mixer and let the dough rest for 10 minutes. Turn the mixer back on and knead on low for another 5 minutes.

✦ Meanwhile, prepare a loaf pan (see pages 180–183).

✦ Flour your hands and move the dough to a cutting board or other work area. Using your fingers and palms, work the dough into a rectangle about 12 x 9 inches. Then roll the dough into a log and fit it into the prepared pan.

✦ Spritz the top of the loaf with Baker's Joy or some vegetable oil, cover with plastic wrap, and stash in a warm place (or the refrigerator, if you have the time). Let it rise until the dough reaches over the top of the pan and slowly bounces back when poked, about 4 hours at room temperature.

✦ Using a serrated bread knife, slash the top of the dough diagonally 3 or 4 times, then place the pan on a rack in position B in a cold oven. Turn the oven on and set the temperature to 400°F. Bake until an instant-read thermometer inserted into the center of the loaf reads 190°F, about 50 minutes.

✦ Allow to cool in the pan for 5 minutes, and then turn the loaf out onto a cooling rack to cool thoroughly before slicing. Since it contains buttermilk and therefore vitamin C, this loaf will keep—tightly wrapped—for 2 weeks at room temperature.

Yield: One loaf

Notes:

I like the ones with a lip to catch any overflow.

Hardware:

Digital scale

Dry measuring cups

Wet measuring cups

Measuring spoons

Stand mixer with dough hook

Large bowl

Towel

Plastic wrap

Baking sheet

Serrated bread knife

Instant-read thermometer

Cooling rack

Notes:

Whole Wheat Morning-After Bread

A couple of room mates and I used to make this in college. It combined whatever beer we could find the morning after a bender—excuse me, all-night study session—and the first cup of coffee of the day. We'd boost the flour from the pizzeria across the street where we all worked.

THE SPONGE

INGREDIENT	Weight		Volume	Count	Prep
Coffee	227 g	8 oz	1 cup		hot and strong
Beer	227 g	8 oz	1 cup		the darker the better
Whole wheat flour	360 g	12¾	2½ cups		
Instant yeast	7g	¼ oz	2½ teaspoons	1 envelope	

THE FINAL DOUGH

INGREDIENT	Weight		Volume	Count	Prep
Bread flour	405 g	14½ oz	3 cups		
Kosher salt	7g	¼ oz	1 tablespoon		

THE EXTRAS

INGREDIENT	Weight	Volume	Count	Prep
Wheat germ		1 tablespoon		optional
Vegetable oil for the rising bowl and the pan				

✦ ✦ ✦ ✦ ✦

✦ In the work bowl of an electric stand mixer, stir together all of the ingredients for the Sponge. Stir into a paste, cover, and leave alone at room temperature for 30 minutes.

✦ To make the final dough, place the work bowl containing the sponge onto the mixing stand and fit the mixer with the dough hook. Turn the mixer on to Stir (the lowest speed), add the bread flour and salt, and knead into smooth dough.

✦ Move the dough to a bowl that's been lubricated with vegetable oil. Cover with a clean kitchen towel and leave to rise until doubled.

✦ Fold down, then cover and let rise until doubled again.

✦ Fold down again and shape into an oblong loaf (round's okay, too).

✦ Move to a lightly greased baking sheet and sprinkle with the wheat germ, if using.

✦ Cover yet again and allow to rise for another 30 minutes.

✦ Position an oven rack to B and preheat the oven to 400°F.

✦ Slash the loaf 4 times diagonally, cutting no more than 1/3 inch deep.

✦ Bake 20 to 30 minutes or until loaf reaches an internal temp of 190°F.

✦ Remove to a rack to cool completely before slicing.

Yield: One oblong or round loaf

Notes:

Hardware:

Digital scale

Dry measuring cups

Wet measuring cups

Measuring spoons

Small saucepan

Whisk

Stand mixer with
dough hook

Large stainless steel or glass bowl

Clean kitchen towel

Sheet pan

Cutting board

Bread knife

Notes:

Focaccia

This is a rather simplified version of the famed Italian cornbread, which is—to my mind—the most versatile non-traditional loaf bread on the planet. It can be split for sandwiches, grilled, cut into strips and fried, or even rolled thin and used as a pizza dough.

THE LIQUID

INGREDIENT	Weight		Volume	Count	Prep
Water		13 oz	1⅔ cups		divided
Olive oil	56 g	2 oz	4 tablespoons		

THE DRY GOODS

INGREDIENT	Weight		Volume	Count	Prep
Cornmeal	75 g	2⅝ oz	½ cup		
Bread flour	879 g	1lb 15 oz	6½ cups		
Instant yeast	13 g	½ oz	1½ tablespoons		
Salt	24 g	1 oz	4 teaspoons		

THE EXTRAS

Additional olive oil for lubing and drizzling ✦ Kosher salt for sprinkling

✦ ✦ ✦ ✦ ✦

✦ Bring 1 cup of the water to a boil in a small saucepan. Slowly whisk in the cornmeal and cock, whisking constantly, until the mixture thickens, 30 seconds to 1 minute.

✦ Add the remaining liquid to the saucepan, stir to combine, and heat to a temperature of 110° to 115°F.

✦ Pour the liquid mixture into the work bowl of an electric stand mixer fitted with a dough hook.

✦ Add half of the flour and the yeast. Mix with the dough hook on low speed until combined and then allow the dough to rest for 10 minutes.

✦ Again working at low speed, gradually stir in the remaining flour and the salt until the dough comes cleanly from the side of the bowl.

✦ Knead for 5 to 7 minutes if using the mixer, 8 to 10 minutes if kneading by hand.

✦ Turn the dough out, fold over 4 or 5 times, then place in a well-oiled bowl, cover with a clean kitchen towel, and set aside to rise for 1 hour.

✦ Spread the dough out flat on a sheet pan, cover with a clean kitchen towel, and allow it to rise for 1½ hours.

✦ Place an oven rack in position B and preheat the oven to 450°F.

✦ Ten minutes before placing the dough in the oven, dimple it with your fingertips, then drizzle the dough liberally with olive oil and sprinkle it with kosher salt.

✦ Bake for 20 to 25 minutes, or until golden brown.

✦ Remove from the pan to a cutting board. Cut into serving-size rectangles and serve warm.

Yield: One 11½ x 16-inch focaccia

Notes:

In other words, poke shallow depressions—but not holes—in the dough.

The Egg Foam Method

By and large, baking is about blowing bubbles. The bubbles built by the mixing methods discussed so far have been based on flour protein, starch, solid fat and, of course, water. These bubbles have been blown up by water vapor and by the gases created by chemical and biological forces. However, the bubbles in egg white foams are blown up by the conversion of water into vapor and the expansion of air—that's it. Everything that is needed to create the millions of flexible balloons that make up a meringue, soufflé, or angel food cake exists in the miraculous matrix of the egg white.

The Egg Foam Method

1. Separate fresh eggs while cold via the 3-bowl method (see page 265), placing the whites in a clean metal bowl.

2. Using a clean balloon whisk (manned by a very stout arm) or an electric hand mixer (use a stand mixer if you must) beat the egg whites on low speed until they are foamy.

3. Add your acid, such as cream of tartar.

4. Increase the mixer speed to high (at least 75 percent power) and move the beaters around the bowl in alternating directions.

5. When the foam becomes opaque, start sprinkling in the sugar (if called for) slowly so as few bubbles as possible are damaged.

6. Continue beating until the foam reaches the desired stage: soft, medium, or stiff peaks.

If the recipe doesn't call for water, feel free to replace one of the egg whites with warm water. This will not only help the foam come together faster and keep it plastic after beating; it will also help to get the temperature up quicker than having it wait around on the counter.

Properly prepared, egg white applications are impossibly light, airy, delicate, and in some cases, completely fat free. Since they are mostly air, egg foam batters have a nasty reputation as temperamental, troublesome, and downright mean. Like Dr. Seuss's Pale Green Pants with Nobody Inside Them, I believe that egg foams are simply misunderstood. Learn what they need and you'll be in possession of the roadmap to a wide range of delicious culinary destinations.

Whether it's sudsing your hair, topping your beer, or floating in a tide pool on a windy day at the beach, a foam is simply a collection of small bubbles. And a bubble is but a pocket of gas surrounded by a thin layer of liquid, which is surrounded by more air. Now, no pure liquid can produce bubbles. Why not? Surface tension. Take water, for example. Water molecules are so attracted to each other that when faced with an alien environment, like air, any body of water will round up its covered wagons and shape itself so as to expose as few of its molecules to that environment as possible. It just so happens that the shape that exposes the fewest molecules is a sphere.

So, if we wish to make water (and an egg white contains little else) into a foam we must break the tension by adding molecules that can infiltrate the water and get it to loosen up a little…soap for instance.

Egg whites foam fabulously because their water content already contains the necessary foaming agents in the form of proteins. In fact, when it comes to building egg foams, you can usually replace 25 percent of the egg whites with H_2O at a rate of one tablespoon per white. There'll still be plenty of bubble-blowing proteins to go around but the lighter mass will foam a lot faster. The resulting foam will be more flexible and less prone to over-beating.

Eggs are unique because they can be coaxed into foams of such volume and stability that Mr. Bubble would blush. In fact, nothing in nature is as good at foam building as eggs, especially egg whites, which can be whipped into a stable foam nearly eight times their original volume. Whole eggs can be whipped to about four times their original volume (though it can be tricky), and egg yolks one to one-and-a-half times their original volume (don't feel bad for them, they make up for it in other ways).

Creating a foam is a lot like calling a square dance. There are a bunch of proteins and water that need to move together in a very specific arrangement and it's up to you to coax them into it. Now, let's say you have four egg whites in a bowl. How did they get there, you ask? Okay, let's talk about separating for a second.

Here's how I do it:

Why the center cup? Well, sooner or later you're going to have an accident and you're going to break a yolk and contaminate the white. The quarantine container prevents mass contamination. Is a little yolk that big of a deal? Yes, no . . . and maybe. Keep reading.

Fresh eggs separate easier than old, and cold separate easier than warm. But egg whites whip easier at room temperature, so either allow the whites to sit for half an hour or so on the counter or float their container in warm water for a few minutes.

Adding a small amount of acid is smart. Why? Because an acid such as cream of tartar can work to denature proteins so that the beginning foam (the proto-foam) forms faster. Acid also helps prevent over-coagulation of the proteins once the bubbles are blown and that means you've got a little more leeway on the back end of construction.

Although soufflés can be mixed with stand mixers, I always find that there's a little pool of egg white left in the bottom of the bowl. Since you can create chaotic currents with a hand mixer, that's how I go. I used to tout a KitchenAid, but lately the Cuisinarts seem to have a bit more gumption if you get my meaning.

Okay, we make foam now.

When you're ready to beat, do so on a low mixer speed until the whites form a big, wet proto-foam, then put the spurs to 'em. In just a few minutes you'll have a foam that holds the tracks and trails left by the beaters. At this point, start checking your peaks.

Turn the mixer off and hold it so that the beater shafts are straight down. Push the beaters to the bottom of the bowl, and slowly pull them straight out. When the beaters are clear of the mass, turn the mixer so that the beaters are pointed straight up.

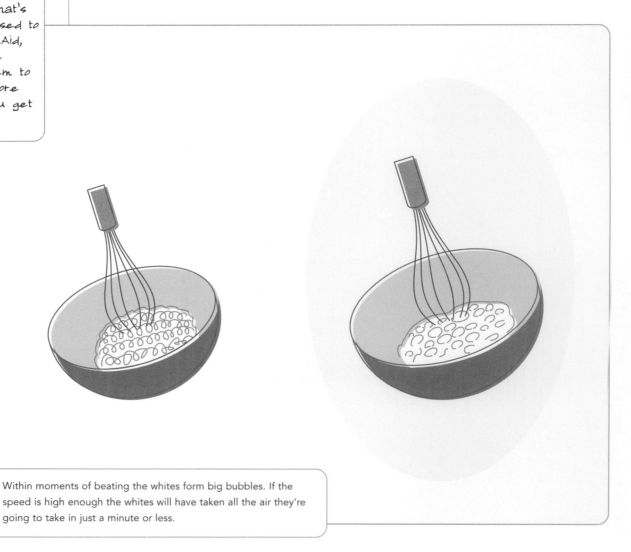

Within moments of beating the whites form big bubbles. If the speed is high enough the whites will have taken all the air they're going to take in just a minute or less.

✦ If the beaters sport twin peaks that just barely fall over when the beaters are tapped, you've got firm peaks and you're good to go. If you can see that peaks are forming but they're completely flopped over, you have more beating to do.

✦ If you have no peak action at all, you have even more beating to do.

✦ If the mess in the bowl looks like a mass of dry cotton floating on water, you've over-beaten. Throw it away and start over.

If you find yourself facing a busted foam, here's what happened: The proteins in the bubble walls squeezed so tight that they squeezed out the moisture they were holding on to and then collapsed on themselves like a zillion little dying suns. So you're left with a bunch of knotted albumin protein floating on water. And there's no cure…luckily eggs are cheap.

Even if you don't beat your eggs to death as described above, you still may have overbeaten your eggs without knowing it. Remember, if you beat the eggs until you have very stiff peaks, odds are good that unless you're working with a sweet meringue that's supported by a lot of sugar, your product will fail to expand in the oven. That's because each of the bubbles in the foam is like a balloon; when it's full, it's full.

Take a balloon and blow it absolutely as full as it can be and tie it off. Now take it out in the sun on a hot day and see how long it takes to burst. Not long, because the walls of the bubble (latex) couldn't expand as the air inside heated. The same thing can happen inside a soufflé or angel food cake. So beat your whites until you have a foam that's stiff and glossy but only until it starts looking dry…beyond there be dragons

For notes on building a sweet foam, see the meringue procedure on the next page.

Hardware:

Digital scale

3 glass or metal bowls for separating eggs

Large metal bowl

Stand mixer, electric hand mixer, or balloon whisk

Dry measuring cups

Wet measuring cups

Measuring spoons

Chef's knife

Cutting board

8- or 9-inch pie plate or pan

Flexible spatula

Notes:

Baked Meringue Pie Crust

You can fill this crust with instant pie filling and top it with whipped cream (see pages 310–311) and eyes will still roll back in heads. I think this is a darned good example of what egg whites can do if they put their foam into it. And don't worry, the fat from the nuts won't make the meringue fall—it will anyway.

THE FOAM

INGREDIENT	Weight		Volume	Count	Prep
Egg whites	60 g	2 oz		2 large	
Cream of tartar	< 2 g	< 1/8 oz	1/8 teaspoon		
Salt	< 2 g	< 1/8 oz	1/8 teaspoon		
Vanilla extract	< 2 g	< 1/8 oz	1/2 teaspoon		

THE REINFORCEMENT

INGREDIENT	Weight		Volume	Count	Prep
Sugar	112 g	4 oz	1/2 cup		

THE EXTRAS

INGREDIENT	Weight		Volume	Count	Prep
Walnuts or pecans	85 g	3 oz	1 cup		finely chopped
Vegetable oil or shortening for the pan					

✦ ✦ ✦ ✦ ✦

✦ Place an oven rack in position C. Preheat the oven to 300°F and lubricate an 8- or 9-inch pie plate or pan with oil or shortening.

✦ Assemble the meringue mixture via the **EGG FOAM METHOD,** adding the salt with the cream of tartar. Add half the sugar when the foam begins to build, then continue to add the rest of the sugar in smaller batches, whipping until stiff peaks form. Gently fold in the vanilla and nuts.

✦ Use a flexible spatula to mold the meringue into the plate, shaping it like a crust, but don't go over edges.

✦ Bake for 50 minutes, until the meringue is firm and lightly browned.

✦ Allow the meringue to cool completely, at least 2 hours.

✦ Fill with a no-bake filling such as pudding or sliced fresh fruit—anything you'd put in a pre-baked pie shell—and top with whipped cream.

Yield: One 8- to 9-inch pie shell

Note: In warm weather, meringues will get gummy after a few days, so it's best to serve this pie shell within 24 hours. You could freeze it, but I wouldn't. The meringue will lose its seductive chewiness.

Notes:

Spiced Angel Food Cake

Although there is nothing wrong with plain angel food cake, it is rather, well . . . plain. After all, it's mostly air, which is why I think it makes sense to add spices that are on the aromatic side. You know, to flavor all that air.

THE FOAM

INGREDIENT	Weight		Volume	Count	Prep
Egg whites	360 g	12 oz		12 large	at room temp
Water	85 g	3 oz	⅓ cup		warm
Cream of tartar			1½ teaspoons		

THE REINFORCEMENT

INGREDIENT	Weight		Volume	Count	Prep
Sugar	392 g	14 oz	1¾ cups		
Salt	2 g	< ⅛ oz	¼ teaspoon		
Cake flour	128 g	4½ oz	1 cup		sifted
Allspice			¼ teaspoon		
Mace			¼ teaspoon		
Ground cloves			⅛ teaspoon		

✦   ✦   ✦   ✦   ✦

Notes:

✦ Place an oven rack in position C and preheat the oven to 350°F.

✦ In a food processor, spin sugar about 2 minutes until it is superfine and remove it to a separate bowl. Return half of the sugar, and all of the salt, cake flour, and spices to the food processor and sift together, setting the remaining sugar aside.

✦ Assemble the egg whites and water via the **EGG FOAM METHOD,** adding the cream of tartar as directed.

✦ Slowly sift the reserved sugar over the opaque foam as directed.

✦ Once you have achieved medium peaks, sift in enough of the flour mixture to dust the top of the foam. Fold in gently using a spatula (see page 307). Continue until all of the flour mixture is incorporated.

✦ Carefully spoon mixture into an ungreased tube pan. Bake for 35 minutes before checking for doneness with a toothpick. (When inserted halfway between the inner and outer wall, the toothpick should come out dry.)

✦ Cool upside down on cooling rack for at least an hour before removing from pan.

Yield: 10 to 12 servings

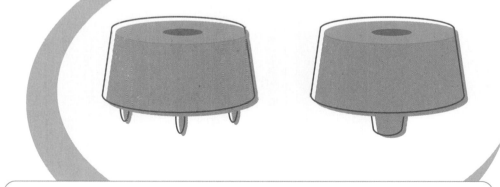

Angel food cake must cool in a suspended state. That's why an angel food cake pan will often have those little feet on the rim. Others have a center tube that's taller than the sides. If your tube pan has neither, you can suspend the pan on a drinking glass for support.

Soufflé Technique

Along with angel food cakes and meringues, tossing soufflés is a cheap, delicious way to hone your egg foam skills. Embedded into this technical review is the recipe for a basic, no-nonsense cheese soufflé that has never let me down. I even made it with Tabasco-spiked Velveeta cheese once and it still passed muster. See page 280 for the complete recipe.

soufflé is French for "puffed up."

Although the soufflé has long been seen as the big, bad boogie man of the culinary canon, the truth is that soufflés are neither complex nor difficult. All they require of the cook is a little understanding and some careful steering around a couple of procedural potholes. No big deal. And there's the payoff, which in my mind is substantial.

Part of what makes soufflés fascinating is that they are binary systems composed of two completely unrelated components…sort of a yin and yang, if you will. One side of the equation is a light, relatively fragile egg-white foam, which is responsible for growing to a couple of times its original size while carrying a thick, heavy roux-based sauce—or base—on its back. This base in turn is expected to deliver 100 percent of the final construct's flavor while taking up a small fraction of its mass.

What Makes This All Work:

✦ Choosing the right vessel and prepping it properly.

✦ A properly constructed foam.

✦ Getting enough flavor into the roux base.

✦ Proper joining of the two components.

Pan Selection

Soufflés may come in many sizes but they only come in one shape: round. Now just about any tall-sided ceramic vessel will suffice for a soufflé, but none does a better job than an actual soufflé dish that has some design modifications especially suited to the task at hand.

Unlike many baking dishes that have slightly sloping sides, the soufflé dish sports a 90-degree angle from floor to wall. These straight sides promote climb. But look up at the top. About three-quarters of an inch from the summit is a lip that flares slightly outward. If the soufflé breaches the top it will spread out rather than climb straight skyward— a trajectory that would probably lead to early collapse. Notice the fluted exterior. It actually

Notes:

soufflé dishes can be used for anything. I bake banana pudding in mine.

Notes:

increases the outer surface area so that heat can be absorbed quicker. And then there's the unglazed bottom. Some say the lack of glaze speeds the absorption of heat from below…I'm not saying I don't buy that, but I'm not saying that I do either. My favorite size is a number seven, which holds right at 1½ quarts, perfect for a five to six egg soufflé serving three to four people.

Pan Prep

Eggs contain a lot of protein, and protein is notorious for sticking to the surfaces of cookware. So our collective instinct is to lube those sides up with some butter. But here's the catch: soufflés rise best when they can get some traction. The answer is to lube with butter, then thoroughly dust with something dry. Although flour is traditional, I usually use finely grated Parmesan cheese for savory soufflés and cocoa powder for sweet.

Here's How It Works:

✦ Rub the entire interior of the soufflé dish with butter. Take special care to cover the corners where the sides meet the floor. Dump in a couple tablespoons of grated Parmesan and tightly cover with plastic wrap. (A fair amount of that cheese will be recovered before the soufflé mixture goes in.)

✦ Roll and shake the soufflé dish so that the cheese has a chance to make plenty of contact with the butter.

✦ Stash the soufflé dish in the refrigerator for 10 minutes.

The Base

Traditional savory soufflés depend on a white sauce as a base. And by white sauce we mean a milk-based sauce thickened by a roux. A few friendly words about roux: A roux is simply a cooked paste composed of an equal amount (by weight) of all-purpose flour suspended in butter. Although you could make roux in any amount, here's a formula that makes enough for one 5-to-6-egg soufflé:

Roux:

INGREDIENT	Weight		Volume
All-purpose flour	21 g	¾ oz	3 tablespoons
Butter	28 g	1 oz	2 tablespoons. melted

Yes, I know I said "equal amounts by weight," but remember that butter is about 15 percent water and other stuff besides fat and we're going to cook the water out before introducing the flour to the party.

To Assemble the Roux:

✦ Melt butter over medium heat.

✦ When the butter stops foaming, all the water will have evaporated. Sprinkle on the flour and whisk to combine.

The goal here is to coat each grain of flour with fat and then cook that flour until the raw starchy flavor is gone. Why bother with coating the flour with fat? Because if each grain is coated with fat, the flour won't turn into a bunch of gooey lumps upon introduction to the hot liquid it's going to be thickening. Water is also the reason for waiting to add the flour to the butter until after the bubbling is over. Bubbles mean that water is turning to vapor and as long as it's happening there's water in that pan and as we've already established, water and flour gets you lumps nine times out of ten.

✦ Now drop the heat to low and continue whisking and cooking for two minutes. At first the paste will get really tight, then it will loosen up again. This is normal.

✦ In the meantime, combine milk, dry mustard, garlic powder, and salt in a tightly sealable container and shake the beejeebers out of it.

Notes:

The easiest way to do this is cut 3 tablespoons from the stick of butter, rub enough on the soufflé to properly lube the interior, then melt the rest.

Notes:

To Complete the Sauce Base:

✦ Whisk the milk mixture into the roux and turn the heat to high. The wheat starches won't gelatinize completely until they reach a boil. Many is the impatient cook who adds more flour only to find his or her sauce turns to library paste at a boil.

✦ Meanwhile beat the egg yolks in a small bowl until they turn creamy in texture and light yellow in color. This will help to denature some of the proteins and evenly distribute the fat so that the yolks will be a lot less prone to scrambling when they join the hot sauce. Tempering will help too.

✦ Whisk a little bit of the hot mixture (whatever holds onto the whisk) into the beaten egg yolks. Continue in this manner until about a third of the hot mixture has been added to the eggs. Then whisk the egg mixture into the hot mixture.

✦ Once the sauce is constructed, stir in the cheese a handful at a time. Don't do this over direct heat, just take advantage of whatever heat is left inside the pot. Additional heat will only result in a grainy sauce.

✦ Now give it a taste. Pretty strong, huh? That's good, because although a soufflé base makes up only about 40 percent of the soufflé's volume, it has to deliver all the flavor.

The Foam

You'll notice that the recipe calls for one more egg white than yolk That shouldn't be a problem if you've been freezing the whites left over from your forays into custard making (see pages 288–293).

✦ Five room-temperature egg whites and a tablespoon of H_2O go into a clean metal bowl, along with ⅛ teaspoon of cream of tartar.

✦ Now we beat it, starting low, then high as described on pages 266–267.

> Yes, technically you could do this in a glass bowl, but I just don't like the idea of slamming a whisk around in something that's heavy—or breakable.

So Here's Where We Are:

✦ Fresh foam formed? Check

✦ Room temperature base standing by? Check

✦ Prepped soufflé dish on deck, ready to go? Check

Oven...you did remember to turn the oven to 375°F and set a rack to position B, right? Good. And did you make sure there are no racks above this? Double good.

Notes:

I once had a soufflé rise so beautifully that when it was ready to come out of the oven the rack above it was baked into it . . . don't let this happen to you.

Folding

✦ Time to fold the foam and sauce together, so check out the note on Folding on page 307.

✦ Start the fold by rapidly stirring a third of the egg whites into the sauce; then fold in half of the remainder, then the rest.

Prep and Bake

Pour the mixture into the prepped soufflé dish and use your spatula or a plate to smooth the top. Now when it cooks, this soufflé is most likely going to mushroom over as it rises above the edge. That's just natural expansion. You can restrict this outward growth by placing your thumb at the edge of the vessel, pushing down about half an inch and rotating the soufflé so that you basically carve a shallow ditch around the entire perimeter. Believe it or not, this miniscule maneuver will greatly improve the appearance, not to mention the texture of the final dish.

Now straight into the oven—and I do mean straight. And don't forget the pie pan that goes underneath. It'll make things a lot easier in the end. Now set your timer for 35 minutes.

Various sitcoms, movies, comic books and the occasional Jane Austen novel would have us believe that once the soufflé is interred, the house must be evacuated lest vibrations make it fall. This is not the case. I'm not saying that a newly ovened soufflé could handle the seismic activity of a jump-roping hippo, but everyday running around will do no harm.

Notes:

Damage during baking is usually the result of opening the door within 30 minutes of the soufflé going in. Breaching the oven in any way will result in a loss of heat and that may result in a permanently deflated foam. So don't open that door. Just look through the window—that's why it's there.

When Is It Done?

Your soufflé is all puffed and proud and golden brown, but what about the inside? The only way to know for sure is to punch a wee little hole into it. Just take a paring knife, find a convenient crack or fold in the top, and plunge straight down. It should be ooey and gooey down there, but if you see a lot of liquid running around, slide it right back in for another 5 minutes.

How to Serve a Soufflé

"Carve" the soufflé with a big metal spoon, making sure to plunge the device all the way to the bottom before spooning out the mixture. The goal in serving is to make sure that every diner gets a cross section from the crusty top all the way to the gooey bottom. Gooey bottom? That doesn't really sound very good does it? Well, in this case, it is.

Why Fat Messes Up Egg Foams

LET'S SAY YOU GET DRAGGED TO A PERFORMANCE OF *RIVERDANCE*.
When all those quick-steppin' freaks link up arms and form an annoying human chain, they are imitating the proteins in an egg-white foam. That is, they hook up with each other to create a molecular grip onto which water can cling.

Now, let's also say that you have that pocket watch from *The Twilight Zone*. You know, the one that lets you stop time? Let's say you use the watch to stop time, run down to the hardware store, and buy some axle grease. Then let's say that you went back to the theater and smeared the grease all over the dancers' arms, returned to your seat, and restarted time.

When all that fancy footwork started up again, I'm betting the scene would look something like this:

Besides being a great moment in show business history, this would also be a darned fine example of what happens when egg yolk gets into the whites. The fats in the yolk lubricate the junctions between the proteins. They can't hold together anymore so the walls of the bubbles collapse.

This is also why beating egg whites in plastic is a no-no. Fats and plastics share common molecular structures. In fact, they often bond together so tightly that even a serious washing can't drive them apart. As a result, plastic utensils often harbor enough fat molecules to crash a batch of whites.

Now, here's a codicil. The truth is, if you have a strong mixer, a foam can be erected with fat in the equation. I recently made a four egg-white meringue that included half a yolk and had no problem. I had to turn the mixer all the way up to its highest speed and it took several more minutes, but after I added sugar and some water to the equation, it did achieve stiff peaks and was a stable foam.

Hardware:

Digital scale

Wet measuring cups

Dry measuring cups

Measuring spoons

Food processor

Box grater

Saucier or sauce pan

Small bowl or teacup

Stand mixer, electric hand mixer, or balloon whisk

Large metal bowl ◄

Hand mixer

1.5-quart soufflé dish

Heavy duty plastic wrap

Disposable pie tin

Large rubber spatula

Notes:

Very, very clean.

Also very clean.

Cheesy Soufflé

Although this rather classic soufflé features sharp cheddar, the truth is that you can make it from almost any cheese, including Velveeta. Smoked Gouda is also a darned fine addition. Oh, and as for the Parmesan that's used on the soufflé dish itself, that can be replaced with any cheese that's hard enough to grate: Pecorino Romano comes to mind, as does aged Gruyère.

THE SAUCE

INGREDIENT	Weight		Volume	Count	Prep
Unsalted butter	43 g	1½ oz	3 tablespoons		
All-purpose flour	21 g	¾ oz	3 tablespoons		
Dry mustard			1 teaspoon		
Garlic powder			½ teaspoon		
Kosher salt			⅛ teaspoon		
Milk	312 g	11 oz	1⅓ cups		hot
Egg yolks	80 g	3 oz		4 large	
Sharp cheddar cheese	170 g	6 oz	1½ cups		grated

THE EGG FOAM

INGREDIENT	Weight		Volume	Count	Prep
Egg whites	150 g	5 oz		5 large	
Water	14 g	½ oz	1 tablespoon		
Cream of tartar			½ teaspoon		

THE EXTRAS

INGREDIENT	Weight		Volume	Count	Prep
Grated Parmesan cheese	50 g	1¾ oz	½ cup		*(not shredded)*

✦   ✦   ✦   ✦   ✦

✦ Place an oven rack in position B and remove any other racks from higher positions. Preheat the oven to 375°F.

✦ Peel back the wrapper on the butter and, holding it like a really big lipstick, thoroughly grease the interior of the soufflé dish—be especially careful to get into the corners. Reserve the remaining butter.

✦ Add the Parmesan and cover the dish tightly with plastic wrap. Shake to coat the entire interior of the dish with cheese, reserving any excess. Place the dish in the refrigerator.

✦ In a small saucepan, heat the remaining butter. Allow all of the water to cook out.

✦ Add the flour to the butter and whisk to combine. Reduce the heat to low and continue to cook for 2 minutes.

✦ Combine the heated milk, dry mustard, garlic powder, and salt in a tightly covered container and shake well to mix. Whisk the milk mixture into the roux and return the heat to high. As soon as the mixture comes to a boil, remove the pan from the heat.

✦ In another bowl, beat the egg yolks until light in color and slightly thickened. Temper the yolks into the milk mixture (see pages 294–295), constantly whisking while doing so. Remove the pan from the heat and add the cheese. Whisk until incorporated.

✦ In one more bowl, beat the egg whites, water, and cream of tartar via the **EGG FOAM METHOD,** to create glossy and firm peaks.

✦ Add ¼ of the egg foam mixture to the base. Continue to add the foam by thirds, folding very gently.

✦ Pour the mixture into the soufflé dish to ½ inch from the top. Place the soufflé dish in the oven on a disposable pie tin and bake for about 35 minutes, or until the soufflé has risen and is browned on top.

✦ Serve immediately.

Yield: Serves 4 to 6

THE TRUTH ABOUT COPPER BOWLS

✦   ✦   ✦

Q: Are copper bowls really better for beating egg whites?

A: Yep.

Q: Could you go into a little more detail?

A: Sure. When you beat egg whites in a copper bowl, copper ions tangle up with a protein called conalbumin in the egg white. Since the new complex that is formed is more stable than the protein is on its own, foams beat in copper are harder to overbeat and they stay fluffy longer. Because of this, cream of tartar need not be added to egg whites whipped in copper.

A: Is it worth spending a hundred bucks on a copper bowl?

Q: Not if you ask me. Good technique applied to egg whites in a stainless bowl will yield equally positive results, especially if cream of tartar is added. Besides, copper is toxic and a beast to clean.

Hardware:

Digital scale

Wet measuring cups

Dry measuring cups

Measuring spoons

Food processor

Box grater

Saucier or sauce pan

Small bowl or teacup

Stand mixer, electric hand mixer, or balloon whisk

Large metal bowl

Hand mixer

1.5-quart soufflé dish

Heavy duty plastic wrap

Disposable pie tin

Large rubber spatula

Notes:

Very, very clean.

Also very clean.

If I find out you've used some lame instant grits or anything other than stone-ground grits, I will hunt you down . . . I will, I swear it!

Gritty Soufflé

Man, do I love this soufflé. Served alongside grilled lamb chops . . . spare me the oysters, this is the food of love. Believe it or not, this also makes for pretty good leftovers the next day. Many is the breakfast that's found me chomping on reheated last night's soufflé. The secret, of course, is the grits. They swell with water and guard it jealously so you don't have to add a lot of water to reheat. A little cream wouldn't be bad though.

THE SAUCE

INGREDIENT	Weight		Volume	Count	Prep
Unsalted butter	57 g	2 oz	4 tablespoons		
Water	454 g	16 oz	2 cups		
Stone-ground grits	64 g	2¼ oz	½ cup		
Garlic				2 cloves	minced
Salt	< 2 g	< ⅛ oz	⅛ teaspoon		
Egg yolks	60 g	2⅛ oz		3 large	
Cheddar cheese	170 g	6 oz	1½ cups		grated
Cayenne pepper			¼ teaspoon		

THE EGG FOAM

INGREDIENT	Weight		Volume	Count	Prep
Egg whites	120 g	4 oz		4 large	
Warm water	14 g	½ oz	1 tablespoon		
Cream of tartar			¼ teaspoon		

THE EXTRAS

INGREDIENT	Weight		Volume	Count	Prep
Grated Parmesan cheese	50 g	1¾ oz	½ cup		(not shredded)

✦ ✦ ✦ ✦ ✦

✦ Place an oven rack in position B and remove any other racks from higher positions. Preheat the oven to 375°F.

✦ Peel back the wrapper on the butter and, holding it like a really big lipstick, thoroughly grease the interior of the soufflé dish—be especially careful to get into the corners. Reserve the remaining butter.

✦ Add the Parmesan and cover the dish tightly with plastic wrap. Shake to coat the entire interior of the dish with cheese, reserving any excess. Place the dish in the refrigerator.

✦ Place the remaining butter, water, grits, garlic, and salt in a medium saucepan over medium heat and bring to a boil. Reduce the heat to low and simmer, partially covered, until the grits are almost done, 10 to 15 minutes depending on the coarseness of the grits.

✦ Place the egg yolks in a metal bowl and beat until light in color and slightly thickened.

✦ Remove the grits from the heat and temper the grits into the yolks. Then add the cheddar, reserved Parmesan, and cayenne.

✦ Beat the egg whites, water, and cream of tartar via the **EGG FOAM METHOD,** to create medium peaks.

✦ Remove the prepared soufflé dish from the refrigerator.

✦ Add ¼ of the egg foam mixture to the base. Continue to add the foam by thirds, folding very gently. Turn the batter into the prepared dish. Place the soufflé dish in the oven on a disposable pie tin and bake for about 45 minutes, or until the soufflé has risen and is browned on top.

✦ Serve immediately.

Yield: Serves 4 to 6

Notes:

See pages 294–295.

Hardware:

Digital scale

Wet measuring cups

Dry measuring cups

Measuring spoons

Chef's knife

Cutting board

Heavy-duty plastic wrap

Heavy wide skillet or frying pan

Paper towels

Small bowl or teacup for separating the eggs

Large metal bowl

Hand mixer

1.5-quart soufflé dish

Heavy-duty plastic wrap

Disposable pie tin

Large rubber spatula

Notes:

Very, very clean.

Also very clean.

You'll notice there is no salt. Believe me, with a can of condensed soup, there's salt aplenty.

Ridiculously Easy Mushroom Soufflé

If a soufflé base can be composed of any highly flavored and viscous substance, why not use canned soup? We tried cream of celery soup, too, but celery soufflé just doesn't cut it. Ditto cream of asparagus.

THE WET WORKS

INGREDIENT	Weight		Volume	Count	Prep
Unsalted butter	28 g	1 oz	2 tablespoons		
White button or cremini mushrooms			2 cups		sliced thin
Scallions				3	sliced thin, on a diagonal
Egg yolks	60 g	2⅛ oz		3 large	
Condensed cream of mushroom soup			10¾ oz	1 can	not heated
Freshly ground black pepper			¼ teaspoon		

THE EGG FOAM

INGREDIENT	Weight		Volume	Count	Prep
Egg whites	150 g	5 oz		5 large	
Warm water	14 g	½ oz	1 tablespoon		
Cream of tartar			¼ teaspoon		

THE EXTRAS

INGREDIENT	Weight		Volume	Count	Prep
Grated Parmesan cheese	28 g	1 oz	¼ cup		(*not* shredded)

✦ ✦ ✦ ✦ ✦

Notes:

✦ Place an oven rack in position B and remove any other racks from higher positions. Preheat the oven to 375°F.

✦ Peel back the wrapper on the butter and, holding it like a really big lipstick, thoroughly grease the interior of the soufflé dish—be especially careful to get into the corners. Reserve the remaining butter.

✦ Add the Parmesan and cover the dish tightly with plastic wrap. Shake to coat the entire interior of the dish with cheese, reserving any excess. Place the dish in the refrigerator.

✦ Heat the remaining butter in a heavy wide skillet or frying pan over medium heat. Add the mushrooms a handful at a time and sauté them over very high heat. Remove each batch to paper towels as it's done. The goal is to create a crust on the mushrooms. If liquid pools in the pan it's either overcrowded or not hot enough. Quickly sauté the scallions in the remaining butter.

✦ Place the egg yolks in a metal bowl and beat until light in color and slightly thickened.

✦ Place the soup into a large bowl and whisk in the egg yolks. Then stir the mushrooms, green onions, black pepper, and reserved Parmesan into the soup-egg mixture.

✦ Remove the prepared dish from the refrigerator.

✦ Beat the egg whites, water, and cream of tartar via the **EGG FOAM METHOD,** to create medium peaks.

✦ Add ¼ of the egg foam mixture to the base. Continue to add the foam by thirds, folding very gently. Turn the batter into the prepared dish. Place the soufflé dish in the oven on a disposable pie tin and bake for about 50 minutes (don't even think about opening the oven door for the first 40 minutes), or until the soufflé has risen and is browned on top.

✦ Serve immediately.

Yield: Serves 4 to 6

Scallion

The mission: To create a flavored gel in a range of textures from firm and moldable to smooth and spoonable. Unlike other applications in this book, these gels are set and held in place primarily by the coagulative power of egg yolks. Due to the particular nature of the egg, particular procedures must be observed.

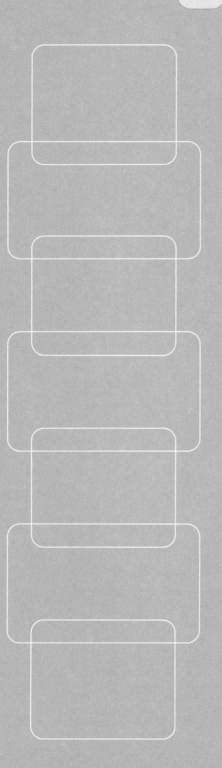

A Tale of Two Custards

The word "custard" comes from the Anglo-French *crustade*, which in turn comes from the Latin *crusta* (from which we also get "crust"). What's most interesting to me is the word's relationship to the Greek *crysta* or "crystal."

An egg can thicken a liquid in two ways. It can emulsify it with fat, as is the case with mayonnaise and hollandaise, or it can thicken by capturing the liquid (particularly dairy products) in a mass of coagulated protein molecules. When the latter is the case we call the result a custard. A surprising number of common culinary applications are indeed custards, including quiche, most puddings and ice creams, curds, various dessert sauces like zabaglione, various savory sauces, cheesecake, and good ole pumpkin pie.

Despite the custard's limitless incarnations, all custards are either stirred or still. Both spring from similar shopping lists, but when it comes to procedures, differences abound.

Behold, two lists of ingredients, each alike in dignity:

INGREDIENT	Weight		Volume	Count	Prep
Milk	227 g	8 oz	1 cup		
Heavy cream	227 g	8 oz	1 cup		
Vanilla extract	14 g	½ oz	1 tablespoon		
Salt			pinch		
Chocolate, semi- or bittersweet	113 g	4 oz			coarsely chopped
Eggs	100 g	3½ oz		2 large	
Egg yolks	40 g	1½ oz		2 large	
Sugar	70 g	2½ oz	⅓ cup		

And

INGREDIENT	Weight		Volume	Count	Prep
Milk	227 g	8 oz	1 cup		
Heavy cream	454 g	16 oz	2 cups		
Vanilla extract	7 g	¼ oz	1½ teaspoons		
Crème de cacao	14 g	½ oz	1 tablespoon		
Cayenne pepper			⅛ teaspoon		
Salt			pinch		
Chocolate, semi- or bittersweet	227 g	8 oz			coarsely chopped
Eggs	300 g	10½ oz		6 large	
Sugar	70 g	2½ oz	⅓ cup		

Notes:

Freeze the left-over egg whites in ice-cube trays. After the whites are frozen, place the frozen cubes into zip-top freezer bags. The frozen whites can be kept up to a year. What to do with them? See Baked Meringue Pie Crust, pages 268–269.

Notes:

When I used the term "stirred" here, I'm not talking about the churning or freezing of the ice cream, but the stirring that goes on during the actual cooking of the custard.

Okay, so you're wondering what's up with the fire powder. Well, it's an Aztec thing. They mixed hot chiles into their cocoa. The Aztecs may have gotten a few things wrong (blood sacrifices, trusting Europeans with guns), but this they nailed. It's a great combo and as much as I know you're going to be tempted to leave it out—just try it, you'll like it. And if you don't, I'll eat your custard.

The first list is for a baked (still) pudding, the second for an ice cream (stirred). And yet, they are very similar: the ice cream has a third more dairy, twice as much chocolate, and a third more eggs than the pudding. The amounts of liquid flavorings are essentially the same and the sugar amount is identical. So although there are differences here, these applications are still next-door neighbors.

Most of the ingredients added to bring the pudding list up to the ice cream list are there because—although ice cream is indeed a custard—when finished, it will be a custard full of ice. One of the reasons there's more dairy in the ice cream is that some of the moisture will be bound up in the custard's protein matrix and some will be turned into ice crystals. Added fat in the form of chocolate and eggs will help to combat the gritty mouth feel that cheap ice creams have. And the added chocolate and the liqueur help make up for the fact that cold greatly turns down the volume on our taste buds. (If you don't dig chocolate-flavored liqueur, go with an equal amount of coffee-flavored liqueur.) The cayenne in the ice cream is just a little thing I'm fond of.

Up to a point, the procedures of assembly are identical. It is a classic custard story:

✦ Combine the cream and the milk in a heavy saucepan, and add the vanilla and other flavorings. Scald over low heat.

✦ Add the chocolate and stir to melt.

✦ Set aside.

✦ Whisk the eggs and any additional yolks until light, then whisk in the sugar. Giving the yolks a good beating helps them get ready to denature and breaks up some of those bonds that keep them so darned self-contained. Think of it as a pre-game warm up.

Many, nay most, custard recipes call for dissolving the sugar in the milk. I don't. I add it to the eggs because the sugar will help prevent curdling. On the other hand, I can find no advantage whatsoever to adding the sugar to the dairy.

✦ Slowly temper (see page 294–295) the chocolate/milk mixture into the egg mixture.

Only at this point, when the custard is essentially built, do the paths diverge.

For the Still Custard:

✦ Place a oven rack at position B and heat the oven to 300°F.

✦ Place six 6-ounce custard cups in a roasting pan or large metal baking pan.

✦ Bring a pot of water to a boil.

✦ Strain the custard through a fine-mesh sieve into the custard cups, filling each cup ¾ full.

✦ Open the oven door and rest the metal pan on it. Pour enough hot water into the pan to come halfway up the sides of the custard cups.

✦ Carefully set the metal pan onto the oven rack and bake for 25 to 30 minutes, until the custards are set but still a bit wobbly. (The internal temperature will be between 160° and 165°F.)

✦ Using tongs, remove the cups from the pan to a kitchen-towel–lined sheet pan. Allow the water in the roasting pan to cool before discarding.

✦ Allow the pudding to cool to ambient room temperature, then refrigerate for a minimum of 2 hours before serving.

SCALDING

Scalding means to bring a milk and cream mixture just to a simmer, then remove the pan from the heat. The scalding of the dairy not only provides the heat to melt the chocolate, it also creates a smoother custard—just don't ask me why. I've spoken with hardcore food scientists and none of them can give me a straight answer. It may have to do with milk proteins, or with the breaking down of sugars (although that's doubtful at these low temperatures), but I think it's...well, heck, I really don't know why, but I do know that it happens.

Straining is necessitated by those pesky chalazae—the little bungee cords of egg white that keep the egg yolk centered in the egg. They don't dissolve, they just get hard and nasty.

Notes:

If you have a candy/fry thermometer, this would be a good time to clamp it on the side of the pot. If you don't have one or just don't want to bother, use your trusty Thermapen instant digital thermometer. (Don't have one? Just drop by www.Thermoworks.com. They have plenty.)

For the Ice Cream:

✦ Pour the mixture back into the pot, place over low heat, and bring to 170°F, stirring often.

✦ Remove from heat and strain through a fime mesh sieve.

✦ Cool to ambient room temperature, then chill in the refrigerator for at least 4 hours, overnight if possible.

✦ Churn according to your ice cream freezer's instructions. When the mixture has increased in size by a third and is very thick, move to an airtight container and harden in the freezer for at least four hours. Now by the way would be the time to stir in any chunks, like miniature pretzels or chocolate-covered cherries, or…well you get the idea. Of course, you could just let it go naked (not everything has to crunch, you know).

Now, I know what you're thinking. If you could settle on one application for making custard, then you could just make a big batch and split it, making pudding from one half and ice cream from the other. Then you'd have a veritable cornucopia of chocolaty pleasure.

I hear ya.

Although you'll sacrifice a little on both ends with this application, odds are good that you and your loved ones will never notice the difference unless you taste samples side by side.

Here's another mystery for you. No one really knows why this aging or "mellowing" process renders better results. Again, it could be a protein thing, or a sugar thing, or a fat thing. Whatever kind of thing it is, it works. Aged ice cream mixtures are always smoother tasting and smoother melting than non-aged mixtures. If you can't wait overnight, at least allow the mixture to get down to refrigerator temperatures. The colder the mix is when it goes into the churn, the faster it will freeze and the smaller the ice crystals inside will be—and therefore, the smoother the ice cream will be.

Almost Best of Both Worlds

INGREDIENT	Weight		Volume	Count	Prep
Milk	454 g	16 oz	2 cups		
Heavy cream	680 g	24 oz	3 cups		
Vanilla extract	28 g	1 oz	2 tablespoons		
Salt	2 g	<$\frac{1}{8}$ oz	$\frac{1}{4}$ teaspoon		
Bittersweet chocolate	340 g	12 oz			coarsely chopped
Eggs	400 g	14 oz		8 large	
Egg yolks	40 g	$1\frac{1}{2}$ oz		2 large	
Sugar	210 g	$7\frac{1}{2}$ oz	1 cup		

✦ Bake half, turn half, have fun.

Of the myriad custard applications that follow, note that the first is a quiche, the classic interpretation of the custard that the French call a *sauce royale*. In this method, whole eggs are mixed with unscalded dairy and the resulting cold mixture is poured over solids and then baked. You'll also find a zabaglione, which also involves no scalding or tempering.

Which should never be confused with a royale with cheese.

Hardware for the Custard:

3 small glass bowls for separating eggs

Digital scale

Wet measuring cups

Measuring spoons

Chef's knife

Cutting board

Large saucepan

Whisk

Wooden spoon

Fine-mesh sieve

Extra Hardware for the Pudding:

Six 6-ounce custard cups or ramekins

Roasting pan or metal baking pan

Tongs

Clean kitchen towel

Half-sheet pan or cookie sheet

Extra Hardware for the Ice Cream:

Airtight container

Ice cream freezer

Speed Kills

MOST COOKS CAN TELL YOU that if eggs overcook, they curdle. That is, they scramble, forming tight curds of over-coagulated proteins. And while this is certainly true, it's by no means the entire story. The real danger to eggs in custards is not being cooked too hot, but too fast. In fact, the more slowly you cook a custard, the better off you are.

The higher the heat you use, the closer the set point (which you want to hit) and the over-coagulation point (which you want to avoid) are to each other. Imagine trying to win a foot race where the finish line is one foot from a brick wall. What are your chances of crossing the finish line and stopping before the wall? If you're going fast, well…too bad. If you're going slow, odds are you'll be fine. And if that weren't enough reason to back down on the heat consider this: the slower you cook eggs, the lower the setting point drops—and that means you have even more room between the finish line and the wall of pain and agony (not to mention rubbery curds).

With this truth in mind, just about every custard recipe known to man contains references to "tempering eggs" and the use of either a double boiler or a bain marie (or hot water bath), all of which are methods for slowing down heat.

Tempering is a method by which beaten eggs can be integrated into a liquid

Brick wall = curdling

Finish Line = proper coagulation

The faster you head toward coagulation, the more likely you are to curdle.

that—under any other circumstances—would be too hot for the eggs to be added to without curdling. The idea is to slowly rather than suddenly increase the temperature of the eggs.

1. While briskly whisking the eggs, slowly drizzle (or spoon, if it's thick) ¼ to ⅓ of the hot mixture into the eggs, then slowly whisk this mixture into the remaining hot mixture. This will not only slow the absorption of heat by the eggs, it will ensure that the proteins are diluted with other substances, such as sugar, and that will prevent them from reaching each other. The physical blocking of proteins by larger sugar molecules explains why a curd is so much easier to make than say, Hollandaise sauce.

2. A double boiler prevents overheating or the rapid heating of stirred custards by isolating the mixture from the high heat of the cook top. Water is turned to steam and steam bathes the bottom of the vessel. Still, steam is hot stuff and near-constant stirring is necessary.

3. A bain marie, or water bath, is used to keep custards from heating too quickly in the oven. Since water cannot rise above 212°F at sea level this ensures that whatever the water is touching won't rise above 212°F, even in a 350°F oven.

Moral: Slow and steady wins the custard race.

Hardware:

Digital scale

Wet measuring cups

Measuring spoons

Chef's knife

Cutting board

Non-stick frying pan

Medium saucepan

Small mixing bowl

Whisk

Wooden spoon

Cooling rack

Notes:

Pancetta, Goat Cheese, and Chive Quiche

What the French call a quiche, we Americans might plainly call a refrigerator pie. One of the best ways I know to use up leftovers, a quiche can consist of any manner of cooked vegetables and meats, with a sauce royale (uncooked custard) poured over the top.

THE DOUGH

INGREDIENT	Weight		Volume	Count	Prep
Basic Pie Dough			1 recipe		see pages 162–165

THE FILLING

INGREDIENT	Weight		Volume	Count	Prep
Pancetta (Italian bacon)		¼ lb			diced
Yellow onion	113 g	4 oz	1 cup	1 medium	chopped
Milk	227 g	8 oz	1 cup		
Heavy cream	227 g	8 oz	1 cup		
Salt			¼ teaspoon		
Freshly ground black pepper			¼ teaspoon		
Eggs	150 g	5¼ oz		3 large	
Fresh goat cheese	113 g	4 oz			
Fresh chives			2 tablespoons		chopped

✦ ✦ ✦ ✦ ✦

✦ Prepare the Basic Pie Dough according to the instructions on pages 162–165.

✦ Place an oven rack in position C and preheat the oven to 425°F. Roll out the dough to fit a 9-inch pie pan, pinch the edges, and discard excess dough. Blind bake the crust for 15 minutes. Set the baked crust aside and turn the oven down to 350 degrees.

✦ Cook the pancetta in a non-stick frying pan just until it starts to get crispy, about 5 minutes. Add the chopped onions to the pan with the pancetta and continue to cook until the onions are translucent and the pancetta is really crispy, about 5 to 7 minutes longer. Set aside.

✦ In a medium-sized saucepan placed over low heat, combine the milk, cream, salt, and pepper and heat to just under a simmer. Remove from the heat and set aside.

✦ In a small bowl, beat the eggs and then temper into the milk mixture (see pages 294–295).

✦ Cover the bottom of the blind-baked piecrust with the pancetta and onions, and then slowly pour in the egg and milk mixture.

✦ Crumble the goat cheese over the mixture in small amounts, dispersing it evenly across the surface, then top with the chopped chives.

✦ Bake for 35 to 40 minutes, or until the internal temperature reaches 160° to 165°F. The center should still be wobbly.

✦ Allow to cool for 15 minutes before serving.

Yield: Serves 6 to 8

Notes:

Hardware:

3 small glass bowls for separating eggs

Digital scale

Wet measuring cups

Measuring spoons

Small saucepan

Medium saucepan

Eight 6-ounce custard cups or ramekins

Kettle

Large stainless-steel bowl with a spout

Whisk

Wooden spoon

Fine-mesh sieve

Roasting pan large enough to accommodate 8 custard cups with at least 1 inch to spare around

Clean kitchen towels

Paring knife

Tongs

Sheet pan

Notes:

Crème Caramel

The caramel will make a sauce when the custards are inverted.

THE CARAMEL

INGREDIENT	Weight		Volume	Count	Prep
Sugar	158 g	5½ oz	¾ cup		
Water	57 g	2 oz	¼ cup		

THE CRÈME

INGREDIENT	Weight		Volume	Count	Prep
Milk	340 g	12 oz	1½ cups		
Half-and-half	227 g	8 oz	1 cup		
Vanilla extract	5 g	¼ oz	1 teaspoon		
Eggs	150 g	5¼ oz		3 large	
Egg yolks	60 g	2 oz		3 large	
Sugar	105 g	3¾ oz	½ cup		

✦ ✦ ✦ ✦ ✦

Place an oven rack in position C and preheat the oven to 350°F.

✦ To make the caramel, place the ¾ cup sugar in a small saucepan over medium heat. Sprinkle the water evenly over the sugar, swirling the pan to combine the sugar and water. Continue swirling until a clear syrup forms, about 3 to 5 minutes. Increase the heat to high and bring to a boil. Cover the pan and boil for 2 minutes. Uncover and continue to cook the syrup until it turns an amber color, 2 to 3 minutes. Quickly pour equal amounts of the syrup into the 8 custard cups.

✦ In a medium saucepan, combine the milk, half-and-half, and vanilla. Bring to a bare simmer over medium-low heat and then immediately remove from the heat and set aside.

Notes:

✦   ✦   ✦   ✦   ✦

✦ Set a kettle of water to boil.

✦ In a small bowl, whisk the eggs and the egg yolks until they have lightened in color. Slowly whisk in the ½ cup sugar and continue whipping until the mixture is slightly thickened.

✦ Temper the eggs into the milk mixture (see page 294–295).

✦ Place a fine-mesh sieve over a glass or stainless-steel bowl with a spout. Pour the milk-egg mixture through the strainer into the bowl.

✦ Place a clean kitchen towel in the roasting pan and put the custard cups on top. Evenly distribute the custard into the custard cups, going short on the first pass.

✦ Open the oven door and rest the pan on it. Pour boiling water into the pan halfway up the sides of the cups and then carefully move the pan into the oven.

✦ Bake for 40 to 50 minutes, or until the custards still wobble slightly when the pan is wiggled. You can also insert a paring knife midway between the edge and the center. If it comes out clean, the custards are done.

✦ Using tongs, remove the cups from the pan to a kitchen towel–lined sheet pan. Allow the water in the roasting pan to cool before discarding. Cool, cover, and chill the custards for at least 4 hours. They will keep, refrigerated, for up to three days.

✦ To serve, dip the cups into warm water and use a knife to loosen edges, then invert them onto serving plates.

Yield: 8 servings

Hardware:

Digital scale

Dry measuring cups

Wet measuring cups

Measuring spoons

Chef's knife

Cutting board

Two medium saucepans

Food processor

Small mixing bowl

Fine-mesh sieve

Large stainless-steel bowl

Whisk

Small clean plastic spritz bottle

9-inch square casserole or baking dish

Notes:

Glass is best because it lets you see the wafers waiting for you in the bottom.

Not part of the original recipe.

Hasty Pudding

This is a nice, simple banana pudding that's faster than the one on the back of the famous box simply because it's not baked. That said, it does taste a heck of a lot better the next day. I should also note that this is one of the oldest recipes in the Brown family chronicles, or at least that's what I'm told. The first time my grandmother made it for me she told me the recipe was written during the Revolutionary War. When I pointed out that Nilla Wafers weren't invented until 1945, she took the entire pudding and gave it away to the minister who lived next door. That's the last time I questioned the authenticity of a dessert—any dessert.

THE CUSTARD

INGREDIENT	Weight		Volume	Count	Prep
Milk	476 g	16 oz	2 cups		
Vanilla extract	5 g	<$\frac{1}{4}$ oz	1 teaspoon		
Sugar	158 g	4$\frac{3}{4}$ oz	$\frac{3}{4}$ cup		
Flour	21 g	$\frac{3}{4}$ oz	3 tablespoons		
Salt			pinch		
Eggs	200 g	7 oz	$\frac{3}{4}$ cup	4 large	

THE BASE

INGREDIENT	Weight	Volume	Count	Prep
Vanilla wafers		about half a box		
Very ripe bananas		3 to 4		thinly sliced

THE EXTRAS

INGREDIENT	Weight		Volume	Count	Prep
Banana liqueur	56 g	2 oz	$\frac{1}{4}$ cup		
Whipping cream	239 g	8$\frac{1}{2}$ oz	1 cup		for topping

So why the flour? A lot of older pudding/custard recipes include starch to prevent curdling and to absorb any extra liquid that might seep out of the final matrix. Flour is a thickener too, but since this mixture never boils, the flour's thickening power can never be fully released. Logic might lead you to believe you'd be better off with cornstarch—and that would be true if you planned to serve the pudding within a few hours, but come the next day your pudding would be soup. That's because egg yolks contain an enzyme that gobbles up starch molecules. And although the eggs cook long enough to be considered cooked, they don't cook long enough to turn off this enzyme.

✦ ✦ ✦ ✦ ✦

✦ In a medium saucepan, bring about an inch of water to a boil, then reduce to a simmer.

✦ Meanwhile, in another medium saucepan, combine the milk and vanilla. Bring to a bare simmer over medium-low heat and then immediately remove from the heat and set aside.

✦ Place the sugar, flour, and salt in the bowl of a food processor, pulsing a few times to combine.

✦ In a small bowl, whisk the eggs until they have lightened in color. Slowly whisk in the sugar-flour mixture and continue whipping until the mixture is slightly thickened.

✦ Temper the egg-sugar mixture into the milk mixture (see pages 294–295) and then strain through a fine-mesh sieve into a stainless-steel mixing bowl.

✦ Set the bowl atop the saucepan over the barely simmering water, making sure the bottom of the bowl does not touch the water.

✦ Whisk constantly for 15 minutes, until the temperature taken with an instant-read thermometer reads between 170° and 175°F. Remove from the heat and set aside.

✦ Line the casserole with the vanilla wafers, and if you're feeling especially frisky, spritz them with banana liqueur. Spread on a thin layer of the custard, then a layer of the bananas. Repeat all the way to the top.

✦ Chill for at least four hours (or preferably overnight) and then top with a layer of whipped cream (see pages 310–311) before serving.

Yield: 8 to 10 servings

Notes:

Wrong way

Right way

One serving if it's serving me.

Hardware:

3 small bowls for separating eggs

Digital scale

Dry measuring cups

Wet measuring cups

Measuring spoons

Chef's knife

Juicer

Fine-mesh sieve

Rasp or citrus zester

Cutting board

Medium saucepan

Medium stainless steel bowl

Whisk

Kitchen spoon

1 pint bowl or container

Plastic wrap

Notes:

Orange Curd

A custard based on citrus juice is usually called a curd. The most common of these is usually made with lemon juice, but orange juice provides a nice change here. And not only is citrus curd shamelessly easy to make, it may actually be the most versatile substance on earth besides duct tape. You can spread it on bread, biscuits, cake, fold it into soufflés, you name it. If you only know how to make one dessert item you might think about making it citrus curd.

THE CUSTARD

INGREDIENT	Weight		Volume	Count	Prep
Egg yolks	100 g	3½ oz		5 large	
Sugar	210 g	7½ oz	1 cup		
Freshly squeezed orange juice	76 g	2⅔ oz	⅓ cup	1 orange	strained
Orange zest			1 tablespoon		
Unsalted butter	113 g	4 oz	8 tablespoons	1 stick	cut into small pieces and chilled

✦ ✦ ✦ ✦ ✦

✦ In a medium saucepan, bring about an inch of water to a boil, then reduce to a simmer.

✦ Meanwhile, combine the egg yolks and sugar in a medium-size stainless-steel bowl. With a whisk, beat the eggs and sugar together thoroughly for at least 4 minutes.

✦ Measure the orange juice. If a single orange has not given you enough juice, add enough cold water to reach 1/3 cup. Add juice and zest to the egg mixture and whisk until smooth.

✦ Set the bowl atop the saucepan over the barely simmering water, making sure the bottom of the bowl does not touch the water.

✦ Continue to whisk the mixture until thickened, about 8 to 12 minutes. The custard should be a light yellow and coat the back of a spoon.

✦ Remove promptly from the heat and stir in the butter a piece at a time, allowing each addition to melt before adding the next.

✦ Remove to a clean container and cover by laying a piece of plastic wrap directly on the surface of the curd. This will keep a skin from forming on the top.

✦ Refrigerate for at least 4 hours prior to serving. The curd will keep in the refrigerator for up to 2 weeks.

Yield: 1 pint

> It's the same trick that keeps guacamole from turning brown.

> This is essential—use a kitchen timer if you're in doubt.

THE ZEST

✦ ✦ ✦

All of the flavor—well, at least 95 percent of the flavor in a citrus fruit—is on the very outer layer of skin. It's called the zest and it's full of oils as opposed to the acid that you find in the juice. You want to make use of that, so you need to peel it off—but you don't want to go too deep because beyond that lies the pith and it tastes, well, dreadful.

You could use a peeler, if you had a really good one, to just bring off the outer surface but then you'd have to chop it up. You could also use a bar grater, which would get down to the zest, but would take it off in these long strings that would probably get caught in your teeth so you'd have to chop them up, too.

The last choice, and my favorite, is a wood rasp that you can get at a hardware store. These things take off just the perfect amount of zest in perfectly small pieces, so no further work is required.

Hardware:

3 small bowls for separating eggs

Digital scale

Dry measuring cups

Wet measuring cups

Large saucepan

Large stainless-steel bowl

Whisk

Notes:

Zabaglione

One of the most elegant desserts of all time when served in a goblet with seasonal fruit, zabaglione is so simple the guy from *Sling Blade* could make it (to top his biscuits, of course). It's a custard, but instead of dairy or fat the liquid held in the egg yolk's net is wine. The French call this *sabayon* and often flavor it with dry white wine and use it as a sauce for fish.

You'll notice that, operationally speaking, there is only one real difference between this and a curd. The butter in a curd is only added after the custard has cooked because the fat in the butter could get in the way of coagulation. Since the wine holds no such threat, it can be whisked in during coagulation. This also helps to cook away some of the "hot" alcohol flavor and it adds to the speed at which this sauce can be tossed together.

THE CUSTARD

INGREDIENT	Weight		Volume	Count	Prep
Egg yolks	120 g	12 oz		6 large	
Sugar	105 g	3¾ oz	½ cup		
Sweet wine			1 cup		

THE EXTRAS

2 to 3 cups fresh berries of your choice for serving

Sauternes, Muscat, Marsala, or Madeira all work well.

✦ ✦ ✦ ✦ ✦

✦ Apply the **STIRRED CUSTARD METHOD** (see pages 290–292), whisking the egg yolks until smooth and slightly lightened in color.

✦ In a large saucepan, bring about an inch of water to a boil, then reduce to a simmer.

✦ Meanwhile, in a large stainless-steel bowl, whisk the egg yolks until they are smooth and slightly lightened in color.

✦ Gradually beat in the sugar, continuing to beat until thick and lemon colored.

✦ Set the bowl atop the saucepan over the barely simmering water, making sure the bottom of the bowl does not touch the water.

✦ Whisk constantly while slowly adding the wine, then continue whisking over the heat until the mixture triples in size, about 10 to 15 minutes.

✦ Remove the pan from the heat and continue whisking slowly for 1 minute to dissipate any extra heat.

✦ Scoop the zabaglione into dessert dishes and scatter fresh berries on top. Serve warm.

Yield: About 6 cups (much of it air), 8 servings

Notes:

Hardware:

Digital scale

Wet measuring cups

Measuring spoons

Stand mixer with whisk attachment

Small saucepan

Balloon whisk

Large stainless-steel bowl

Large saucepan

Paper plate

Notes:

"Blooming" or soaking the gelatin before heating will soften the granules so they'll dissolve more smoothly.

This dessert will appear to set quickly, leaving you to wonder "why wait so long?" I'll tell you why: gelatin can set up to 80 percent in just an hour or so—but the rest takes longer. So it may look like it's set, but it's only mostly set. The mixture will be a good deal tighter after 6 hours than after 2, but not radically so. Still, give the gelatin time to do its thing and your patience will be rewarded.

Zab Mousse

Now let's say that you like the idea of a light, airy, wine-flavored dessert but you want something more like a mousse. No problem. Now that you've seen how easy it is to make zabaglione, all you need are a few additional ingredients.

THE BASE

INGREDIENT	Weight		Volume	Count	Prep
Zabaglione				1 recipe	see page 304

THE FOAM

INGREDIENT	Weight		Volume	Count	Prep
Water	28 g	1 oz	2 tablespoons		cold
Powdered gelatin			1 teaspoon		
Heavy or whipping cream	397 g	14 oz	1¾ cups		cold
Fine sugar	14 g	½ oz	1 tablespoon		
Vanilla extract	5 g	¼ oz	1 teaspoon		

✦ ✦ ✦ ✦ ✦

✦ Prepare the zabaglione, following the recipe on the previous page.

✦ Pour the cold water into a small saucepan and sprinkle the gelatin over the water. Allow the gelatin to bloom for 10 minutes, and then place the pan over low heat, bringing the mixture to 140°F and dissolving the gelatin.

✦ Meanwhile, in the work bowl of an electric stand mixer, combine the cream, sugar, and vanilla extract. With the whisk attachment, whip the cream to stiff peaks.

✦ Temper (see pages 294–295) the dissolved gelatin into the zabaglione and then fold (see page 307) in the whipped cream.

✦ Chill for 6 hours before serving—I suggest vanilla wafers and a really fine wine as accompaniments.

Yield: Eight servings

Folding

THIS IS ONE OF THOSE MYSTERIOUS PROCEDURES that seems crucial to the success of every application in which it rears its ill-defined head.

The verb "fold" refers to the folding of an airy mixture like whipped egg whites much in the way you might fold a napkin or handkerchief…okay, not exactly, but the aim is to fold the mixture over onto itself rather than stir it, whisk it, or beat it—any of which would only deflate the precious bubbles you've built.

Typically this "folding" is achieved by placing the two substances (a foam and a heavier, more viscous substance) together in a large bowl. The folder then plunges a rubber spatula down the middle of the mass to the bottom of the bowl and sweeps up the side, thus flipping half the mass over on itself. The job is made easier by first mixing about a quarter to a third of the foam directly into the base, quickly, to lighten it.

I do my folding in the biggest bowl I have, which is generally twice the size of whatever I'm folding. And instead of a rubber spatula, I use a paper plate. It's more flexible, it's bigger, it makes less of a mess and gets the job done faster than a spatula. But I don't dig down into the middle, I swoop in from the side while rotating the bowl with the other hand.

Hardware:

Digital scale

Wet measuring cups

Measuring spoons

Chef's knife

Cutting board

2 large metal mixing bowls

Large saucepan

Wooden spoon

Rubber or silicone spatula

Small saucepan or
metal measuring cup

Electric hand mixer

Small serving bowls or
martini glasses

Plastic wrap

Notes:

Not baking chocolate—if it doesn't taste good by itself, don't use it here.

Halloween Mousse

Although I always recommend using the best chocolate you can get your hands on, I usually make this mousse with those little bitty bars of Hershey's Special Dark that you're always left with after Halloween. It takes 40 of the little guys. By the way, the stubby French painter Toulouse-Lautrec supposedly invented chocolate mousse—I find that rather hard to believe, but there you have it.

THE WET WORKS

INGREDIENT	Weight		Volume	Count	Prep
Heavy or whipping cream	397 g	14 oz	1$\frac{3}{4}$ cups		
Dark chocolate	340 g	12 oz			chopped fine
Unsalted butter	57 g	2 oz	4 tablespoons	$\frac{1}{2}$ stick	cut into 8 pieces
Coffee liqueur	85 g	3 oz			
Powdered gelatin			1 teaspoon		

✦　✦　✦　✦　✦

✦ Pour 1½ cups of the cream into a large metal mixing bowl and stash it in the freezer.

✦ Place the chocolate, butter, and liqueur in another large metal bowl and melt over a pot of barely simmering water (filled about 1 inch high), stirring constantly.

✦ Remove from the heat while a couple of chunks are still visible. For the next 5 minutes continue to stir occasionally, until the mixture cools to just above body temperature. Set aside.

✦ Place the remaining ¼ cup of the cream in a small saucepan or large metal measuring cup and sprinkle the gelatin over it. Allow the gelatin to bloom for 10 minutes, then dissolve the gelatin by carefully heating and swirling the pan over low heat. Whatever you do, don't let the cream boil or the gelatin's setting power will be greatly reduced.

✦ Stir the mixture into the cooled chocolate and set aside.

✦ Remove the cream from the freezer and use an electric hand mixer to beat the cream to medium peaks.

✦ Before bringing the mousse and the whipped cream together in culinary matrimony, lighten up the chocolate mixture by stirring in just a little bit of the whipped cream—just so the two become a little more alike in texture. Then fold in half of the remaining whipped cream (see page 307). Don't overdo it or you'll beat all that air out of the whipped cream. Then add the remainder of the whipped cream and fold it in. The mixture should remain light and fluffy—if there are still a few flecks of white, it's okay.

✦ Cover the bowl with plastic wrap or spoon the mousse into individual dessert dishes (martini glasses work very nicely for this), cover the dishes, and refrigerate for at least 2 hours.

✦ Garnish with fruit—or not—and serve.

Yield: 8 servings

WHY CHOCOLATE SEIZES

✦　✦　✦

Think of chocolate as a bunch of tiny, dry particles (chocolate solids) suspended in cocoa butter along with some sugar and perhaps vanilla. When melting chocolate, you have to be careful to keep it bone dry. Add a couple of drops of water to melted chocolate and it can seize into a clumpy mass. Add more liquid and eventually the mixture smoothes back out. Why? My friend and hero, food-science guru Shirley Corriher (reading her book *Cookwise* is nearly as good as going to cooking school) explains it thusly:

"Melted chocolate may be a liquid, but it's a dry liquid, meaning little if any water is present. If a little water gets into the mixture, the dry particles stick to it the way that sugar will stick to your spoon if you put it in coffee first. But, put that crystal-crusted spoon into the coffee and there's enough moisture to re-dissolve the mass."

So between the butter and the liqueur in this chocolate mousse recipe there's enough moisture to keep the chocolate fluid.

Whipping Cream

TINY BUBBLES, IN THE MOUSSE MAKES ME HAPPY…makes me feel loose. Okay, so I'm not Don Ho. But like Don I appreciate bubbles—they're everywhere in this book.

But making the bubbles that make up whipped cream is a little different, because unlike egg whites, which inflate best when warm, whipping cream and everything it touches needs to be kept as cold as possible. Why? Because the bubble walls in whipped cream are supported not by protein but by sticky little globules of fat—and unless they're cold, they won't be sticky.

To whip cream you'll need a big metal bowl and a decent mixer. Although we converted to a stand mixer for most applications in this book, this is the one place I'd rather stick with a hand mixer. It can create more turbulence in the bowl and for whipped cream, at least, that's good. Start beating the cream on a relatively low speed, just to whip up some big proto-bubbles. As soon as you get a good Mr. Bubble Bath thing going, crank on the power.

Here's what's going on:

The walls of these proto-bubbles are made up mostly of water and the water soluble materials inside the cream. There are globs of fat drifting around inside the walls and, of course, there's air inside. Now, the more air you whip into the cream, the more bubbles you make. But there's only so much water to go around so eventually you really can't make more bubbles, you can only subdivide the ones you've already made. Eventually these bubbles become so small that the fat globules touch and bingo, you have yourself a stable whipped cream.

What can go wrong, you ask? Well, if you were to continue whipping, the little fats would eventually squeeze out all the water and all the air and you'd find yourself in possession of homemade butter. Good for you. Now go make some toast.

Assuming that you don't make butter, you're still doomed because even if you keep it cold, eventually the water and fat, which aren't very fond of each other even in this microscopic milieu, will separate unless the water is stabilized with a protein mesh like gelatin.

One little teaspoon of gelatin will stabilize all that water, just let it bloom for 10 minutes in cold water, then dissolve it over low heat. Just keep moving it around until you don't see any more grain and be careful not to boil it because that would damage the gelatin.

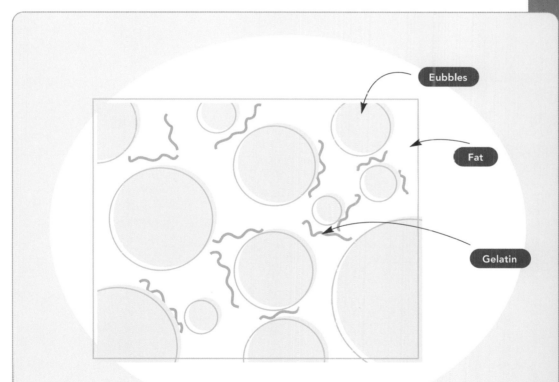

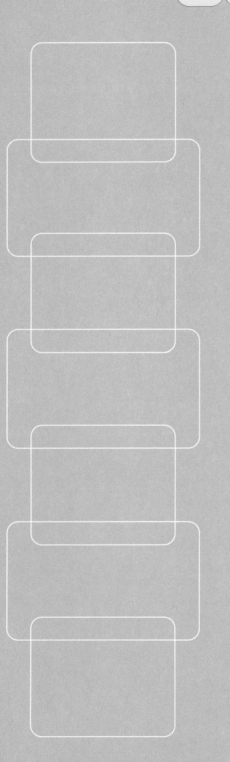

Cheesecake

Cheesecake is a unique and uniquely significant application. Significant because if you can consistently make good ones, fame shall be yours and strangers will bow low when you walk by. Unique in that cheesecake is one of the few devices that dwells in both the custard world and the realm of the cake. It is a custard because it depends on egg protein alone (not flour) for its structure, and yet its texture is derived from the creaming method. Cheesecake is also significant because it is so fine...oh yes, children, it is.

I have pursued cheesecake for a long time now and a couple of years ago felt certain that I had gotten it right. But now I think this one's better—not perfect of course, but better. Shall we begin?

✦ Place an oven rack in position B and preheat the oven to 300°F.

✦ Prep a 9 by 3-inch round cake pan (see pages 180–183).

Crustruction

clever, huh?

Unlike finicky pastry crusts, the pressed or crumb crusts upon which classic cheesecakes are constructed are as much fun to make as they are to eat—and they can be made with almost anything. For instance:

✦ Stale cake crumbs

✦ Crumbled gingersnaps

✦ Crumbled Frosted Flakes

✦ Crumbled Ritz crackers (for savory cheesecake only)

✦ Crumbled vanilla wafers

✦ Crumbled graham crackers

✦ Crumbled Honey Grahams cereal

Ooh, aah.

✦ Beat-up Oreos

A QUICK NOTE ON PANS

✦ ✦ ✦

I've made it a point of public record that I don't care for spring-form pans. I've never seen one in a professional bakery. They break, they fall apart, they leak. I've always relied on Leifheit matte aluminum pans, which don't break, wear out, or leak—ever. Then a friend of mine sent me these fancy new German-made springforms with special bottom lips and a sweet non-stick interior. I used them right away and I have to admit—they leaked just as bad as the cheap ones. I'll still use them, but not for cheesecake.

For now we'll go with the classic taste:

THE CRUST

INGREDIENT	Weight		Volume	Count	Prep
Graham crackers	115 g	4 oz		15 (15 whole crackers, not 15 halves)	
Unsalted butter	115 g	4 oz	8 tablespoons	1 stick	cut into pieces

THE EXTRAS

Unsalted butter or shortening for the pan

Hardware:

9 by 3-inch round aluminum cake pan

Parchment paper

Kitchen shears or scissors

Digital scale

Dry measuring cups

Wet measuring cups

Measuring spoons

Roasting pan

Kitchen towels

Kettle

Electric stand mixer with paddle attachment

Rubber or silicone spatula

Clean kitchen towel

Large mixing bowl

Whisk

Cooling rack

Paring knife

Cake round

Thin, sharp knife

✦ Break up the graham crackers, drop them in the work bowl of a food processor and pulse about 7 times, until there are just a few big pieces left, then add the butter and pulse until the butter disappears and the crumbs slowly fall over when the machine stops.

Use your fingers to pack the crumb mixture into the bottom of the pan, working outward from the middle. You'll end up with a smooth bottom and a pile all around the outer edge. You can turn this pile into the sides of the crust by turning the pan with your thumbs on the outside of the pan and the rest of your fingers on the inside. Press, but don't squeeze too hard or you'll push through the crust.

Bake this blind for 10 minutes, then stick it in the fridge to cool.

Turn the oven down to 250°F and bring a quart of water to boil in a kettle. If you have a really large roasting pan you may need more. You'll want the water to come up to within an inch of the top of the pan. Place a kitchen towel in the bottom of the pan (to prevent slippage).

Custarduction

THE WET WORKS 1

INGREDIENT	Weight		Volume	Count	Prep
Cream cheese	680 g	24 oz	3 cups	three 8-ounce bricks	
Sugar	105 g	3¾ oz	½ cup		

THE WET WORKS 2

INGREDIENT	Weight		Volume	Count	Prep
Eggs	100 g	3½ oz		2 large	
Egg yolks	60 g	2 oz		3 large	
Vanilla extract	14 g	½ oz	1 tablespoon		
Sweetened condensed milk	142 g	5 oz			
Sour cream	227 g	8 oz	1 cup		

That doesn't work as well, does it?

Place the cream cheese with the sugar in the bowl of an electric stand mixer fitted with the paddle attachment. Beat on medium speed until completely smooth and creamy. You'll have to stop and scrape down the bowl sides at least twice—actually, just twice. If tiny little bits

MAKING THE CREAM

✦   ✦   ✦

of cream cheese reach exit velocity and attempt to leave the bowl, simply drape the mixer with a clean kitchen towel.

While the mixer's teaching the cream cheese a thing or two, in a large bowl whisk the eggs and egg yolks together with the vanilla, sweetened condensed milk, and sour cream. (Weighing the sweetened condensed milk is required because it's almost impossible to deal with any other way.)

Now, turn the mixer down to about 30-percent power and slowly—very slowly—drizzle the egg mixture into the cream cheese. Stop halfway through and scrape the bowl, sides and bottom.

Assembly/Baking

As soon as the egg mixture is completely integrated, pour the batter onto the crust. Place the cheesecake in the roasting pan, then open the oven door and rest the roasting pan on it. Fill the roasting pan with the boiling water and place the whole thing in the oven. Close the door and set the timer for an hour. During this hour you will not:

✦ Open the door.

✦ Touch the door.

✦ Mess with the door in any way.

If you want to see what's going on, turn on the light and look through the window. If the window is too dirty to see through, make a note to clean the oven. Why? Because heat energy bounces off clean, shiny walls better than uneven, gunked-up walls. If your oven walls (and door) are nasty, then food can't cook evenly or efficiently. (When I run my oven through its self-clean cycle I usually stick my grill grates in there, too—any built-on gunk will magically vanish, but if they're iron you'll need to re-cure them.)

When the timer goes off, turn the oven off, open the door, and take a look. Doesn't look done, does it? In fact, it looks like soup. Well, that's because the cooking's only half over. Wait 1 minute, then close the door and set the timer for another hour.

MAKING THE CREAM

I used to start this process with the sour cream, then the cream cheese; the idea being that if the bowl and beaters were lubed up a little, the cream cheese wouldn't stick so bad. That was correct but wrong. Lubing isn't what we need. What we need is the pulling, tearing, and slapping that comes from sticking to moving parts. Also, by adding the sour cream to the egg mixture rather than the cream cheese, the sugar doesn't dissolve in the water phase of the sour cream, so it's available to beat little bubbles into the mixture.

Why slowly? Because if you add the egg mixture quickly, it'll sling out of the bowl and all over you.

Although I usually prefer Zyliss spatulas, I have to say that when it comes to scraping down the sides of a KitchenAid mixer, nothing does the trick like a KitchenAid spatula.

THE MOST COMMON CHEESECAKE-MAKING MISTAKES

✦ ✦ ✦

1. Inadequate pan prep.

2. Not whipping the cream cheese enough before integrating the eggs.

3. Not baking long enough.

4. Baking too long.

During this time you will not:

✦ Open the door.

✦ Touch the door.

✦ Mess with the door in any way.

What the heck's goin' on here? The slower we coast across the thermal finish line, the better the odds are that we won't overcook the custard. At this point there's enough residual heat in the oven walls, the water, and the custard itself to finish the job.

When the second hour is up, you may still think the cake's too wobbly for your comfort. But before you curse my name, reactivate the oven, and close the door, remember the scrambled egg axiom: if eggs look done in the pan they'll be overdone on the plate. If you continue to cook this custard until it really looks done, as soon as it cools it's going to start cracking on top. Cracks are the number one sign of over-coagulation.

H₂O   Product   Towel   Pan

Cooling

Cool to room temperature for 1 hour, then chill until the edges of the cake pull away from the side of the pan (3 to 6 hours). Run a knife around the inside edge of the pan just to be sure.

De-Panning

Insert a paring or boning knife straight down between the side of the cake and the parchment and spin the pan while holding the knife still. Fill that roasting pan that you probably left sitting in the oven—or the kitchen sink—with 1 inch of hot water. Set the pan in the water, wait about 5 seconds, then slowly pull the paper straight out.

Move the pan to a towel to catch the water. Place another parchment round on top of the cake, then place a cake round or the bottom of a standard springform pan on top of the

I said I didn't like them, but that doesn't mean I can't try to find another use for them.

cake. Holding the round firmly against the top of the cake, flip the whole thing over. Then simply slide the pan off. Place another cake round or your serving plate upside down on the upturned bottom of the cake and re-invert.

If you've used a springform pan, de-panning is a pretty simple step. Just make sure the side ring is completely unstuck to the pie, slowly undo the buckle, and remove. As for the bottom disk, most folks leave the pie on it for serving. Now the last challenge is...

Cutting

You've heard the expression "like a hot knife through cheesecake"? Okay, it's usually butter, but it might as well be cheesecake, because there's really no other way to cut one. I've heard of people who can cut a cheesecake with dental floss, but I've never been able to get through the crust. So I use a long, thin, sharp knife (a slicer would be good for this) and, if the going gets tough, dip it in hot water every now and then, wipe clean, and cut again.

Eating

No coulis, no sauce, no fruit—okay, maybe fresh fruit—but the way I see it, if the cake is good, why mess it up?

Storing

Wrapped first in plastic wrap then foil, cheesecakes freeze well for up to 3 months.

Crack Kills

If you're none too sure about your timing or your oven's sincerity, you can take out a little crack insurance by adding a tablespoon of cornstarch to the batter when you add the sugar. The starch molecules will actually get in between the egg proteins, preventing them from over-coagulating. No over-coagulating, no cracks. Of course if you do end up with a cake that's cracked on top, there's always whipped cream.

A FINER CRUST

✦ ✦ ✦

I used to advocate a rough and rustic hand-crushed crust. All you had to do was break up the crackers, toss in melted butter, and pack it in the pan. But after playing around with different procedures I've decided that the food processor produces the better crust. Tiny uniform crumbs can compress, which will form a strong but not tough crust, and you can use solid butter, which is a huge advantage because it will then melt in the oven and give up moisture that will help bond the crust and unify it. Melted butter just tends to make things greasy.

Hardware:

Digital scale

Dry measuring cups

Wet measuring cups

Measuring spoons

Small mixing bowl

9 x 3-inch aluminum round cake pan

Chef's knife

Cutting board

Stand mixer with paddle attachment or electric hand mixer

Rubber or silicone spatula

Large mixing bowl

Whisk

Roasting pan

Kettle

Cooling rack

Notes:

I hope that the culinary brave among you will still call it by its true name . . . cheesecake.

Savory Cheesecake

What's in a name? Okay, so here's a darned tasty, savory cheesecake. Although it is absolutely a cheesecake, more than a few folks have suggested that the idea of trout cheesecake is way icky. Fine. Then I'll call it a torte…Smoked Trout Torte. Is that less icky? Good. This is a darned fine appetizer, or can be served alongside a salad as a light lunch or dinner. At room temp, it's even a tasty breakfast. Think of it as a new twist on lox and bagels.

THE CRUST

INGREDIENT	Weight		Volume	Count	Prep
Unsalted butter	85 g	3 oz	6 tablespoons		cut into pieces
Egg whites	30 g	1 oz		1 large	
Bagel chips	85 g	3 oz	2 cups		crushed

THE WET WORKS 1

INGREDIENT	Weight		Volume	Count	Prep
Cream cheese	680 g	24 oz	3 cups	three 8-ounce bricks	room temp
Cornstarch	30 g	1 oz	3 tablespoons		
Salt	6 g	<$\frac{1}{4}$ oz	1 teaspoon		
Sour cream	113 g	4 oz	$\frac{1}{2}$ cup		

THE WET WORKS 2

INGREDIENT	Weight		Volume	Count	Prep
Eggs	100 g	3$\frac{1}{2}$ oz		2 large	

THE EXTRAS

INGREDIENT	Weight		Volume	Count	Prep
Smoked trout	170 g	6 oz			diced
Fresh chives			$\frac{1}{3}$ cup		chopped

✦　✦　✦　✦　✦

✦ Place an oven rack in position C and preheat the oven to 350°F.

✦ Prep a 9 x 3-inch aluminum cake pan (see pages 180–183).

✦ Break up the bagel chips, drop them in the work bowl of a food processor, and pulse about 7 times, until there are just a few big pieces left, then add the butter and egg white and pulse until the butter disappears and the crumbs slowly fall over when the machine stops.

✦ Pack the crumb mixture into the bottom of the pan with your fingers, working outward from the middle and then up the sides.

✦ Blind bake for 8 minutes, until the crust starts to brown.

✦ Remove from the oven and let cool. Reduce the oven temperature to 250°F while you make the filling.

✦ In an electric stand mixer fitted with the paddle attachment, blend the cream cheese, corn-starch, salt, and sour cream until smooth.

✦ Add the eggs, one at a time.

✦ Fold in the trout and chives.

✦ Prepare a roasting pan with a water bath as described on page 315. Pour the batter over the cooled crust and bake for 1 hour. Turn the oven off and leave the cake in the oven for an additional hour without opening the door.

✦ Cool on a rack for at least 4 hours. Carefully unmold as described on page 316. Keep refrigerated until ready to serve.

✦ This will keep in the refrigerator for up to 5 days.

Yield: Serves 12 to 16, more as an appetizer

Notes:

As Well As...

At one time I was going to call this last section the "blender" method. All the recipes would be assembled in said appliance and each leavened in the oven by nothing more than the turning of water to steam. But then I realized that to do this would mean leaving out one of the great steam-leavened devices of all time. So I've decided to break ranks with the rest of the book because it simply makes sense to do so—at least when considering… the Curious Case of Crepes, Popovers, and Pâte à Choux.

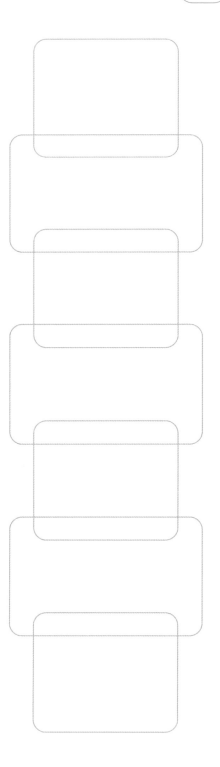

As you'll soon see (if it hadn't crossed your mind already) crepes and popovers are made from nearly identical batters. One stays flat because it's cooked in a flat pan. The other puffs up like a cathedral (with a big empty space inside, to boot) when cooked in a muffin tin or popover pan in a very hot oven. Although constructed from a similar grocery list, Pâte à Choux, or choux paste, differs from crepes or popovers in that its batter is firm and moldable rather than the consistency of heavy cream. That means it can be piped into things like éclairs and cream puffs. How can this be? By cooking this batter on the stove top first, the starch in the flour is gelatinized, thus soaking up and holding on to the water. The resulting dough is every bit as wet as a crepe batter, it's just a lot more viscous—ain't science fun?

Let's make some of each, shall we?

The Blender Method

If a formula contains more liquid than flour, or liquid or melted fat, or no chemical or biological leavening, then the food is leavened only by the steam created during cooking. Two examples that jump to mind (and stomach) are crepes and popovers. I realize that one is expected to climb high while the other is as flat as...well, actually, flatter than a pancake, but the difference isn't in the composition of the batter, it's in the way the batter is handled after it's made. Crepes and popovers are worth talking about because they are extremely versatile, and ridiculously easy to make, mainly because you can make them in a blender.

Why the blender?

- ✦ It's effective.

- ✦ It's fast.

- ✦ It aerates well.

- ✦ It's easier to clean than a mixer bowl.

Blender Batter 1.0

THE WET WORKS

INGREDIENT	Weight		Volume	Count	Prep
Milk	170 g	6 oz	³/₄ cup		
Water	113 g	4 oz	¹/₂ cup		
Eggs	100 g	3¹/₂ oz		2 large	
Salt	3 g	<¹/₈ oz	¹/₂ teaspoon		
Butter	43 g	1¹/₂ oz	3 tablespoons		melted

THE DRY GOODS

INGREDIENT	Weight		Volume	Count	Prep
All-purpose flour	142 g	5 oz	1 cup		

THE EXTRAS

INGREDIENT	Weight		Volume	Count	Prep
Fresh or dry herbs			1 tablespoon		
Additional butter for the crepe pan; butter or meat drippings for the popover pan	76 g	2²/₃ oz	about ¹/₃ cup		

The method:

A blender carafe can be a darned violent place. You don't want the liquid and the flour spinning around in there any longer than necessary or you could end up with a gluten network so tough it would make a gill net blush. So start by placing all the liquids (including fats), eggs, salt, sugar, and flavorings such as vanilla extract in the carafe and blend to combine. Then stop

Hardware:

Digital scale

Dry measuring cups

Wet measuring cups

Measuring spoons

Blender

For the crepes:

8 to 10-inch non-stick pan (see CREPE PANS, page 326)

1 to 2-ounce ladle

Silicone spatula or narrow wooden spatula

Kitchen timer or stopwatch

Basting brush or natural-bristle paintbrush (preferably never used with paint)

For the popovers:

1 six-hole muffin tin or popover pan

Basting brush or natural-bristle paintbrush (preferably never used with paint)

If the fat in question is drippings from a standing rib roast or some other roast critter, the resulting popovers will be what's called Yorkshire pudding. If you use butter, it's just a popover...but that's okay. What you can't do is use shortening or oil. You need butter. Why? You tell me. I'll wait. . . . Right, butter contains milk solids and proteins that add to browning. Not to mention the fact that butter is close to 15 percent water and that water is going to turn to steam, giving the popover even more lift.

Notes:

Do I really need to be telling you that?

the machine, add the flour, put the lid back on and blend again, on low, for no more than 10 seconds total. Note that I say "low" because all the blenders that I have and like only have three settings, including off. If you have one of those blenders that has fifteen buttons, just push one in the middle. By the way, what is the difference between purée and frappe? When measuring, I put the blender carafe on the scale, hit the tare function, and just measure all the Wet Works right into it.

✦ Place the milk, water, eggs, salt, melted butter, and herbs, if using, in the blender and blend smooth, about 10 seconds or so.

✦ Add the flour and blend in bursts until smooth, about 10 more seconds.

✦ If you're planning to make crepes, refrigerate at least one hour. (Skip this step and you'll have big holes and bubbles in your crepes.) The batter will keep for up to 48 hours.

Why "rest" the batter? Keep in mind that when liquidous matter is spun in a blender (and what that goes into a blender doesn't end up "liquidous" in the end?) you get a swirling vortex like this:

Such a vortex is very good for aerating mixtures. So, right after mixing, there are zillions of bubbles in your batter. If you attempt to cook a crepe, those bubbles would blow up and either tear round holes in your crepe or make it lumpy. By resting the batter for an hour, most (but not all) of those bubbles will work themselves out. Also, chilling the batter will make it easier to deal with when it's going into the pan because it won't run around like water.

To make crepes from this batter:

✦ Heat a non-stick skillet over medium heat until a drop of water jumps and sizzles away. Add a teaspoon of butter and, using a clean paint brush (natural bristle only) or pastry brush, paint the bottom and side of the pan with butter.

✦ Pour 1 ounce of batter into the center of the pan and swirl to spread.

✦ Cook for 30 seconds, then flip and cook another 10 seconds then serve—or remove to a rack and cool. Don't stack them until they've cooled, or they will stick together. Once cool, you can stack them and store in sealable plastic bags in the refrigerator for several days. To freeze, place wax paper between the crepes and seal stack in a zip-top freezer bag. (If you plan on serving the crepes right away but need to make several before you can serve, roll the warm crepes and store under plastic wrap and a clean kitchen towel.)

Variation

For sweet crepes, add 2½ tablespoons sugar, 1 teaspoon vanilla extract, and 2 tablespoons of your favorite liqueur to the batter.

✦   ✦   ✦   ✦   ✦

To make popovers from this batter:

A crepe is flat, a popover towering and tall—how can both come from one batter? The afore-mentioned bubbles notwithstanding, it all comes down to the shape of the vessel.

The crepe doesn't rise beyond a millimeter because the steam simply escapes across the wide expanse of the surface. Since the recesses in a popover or muffin pan are relatively narrow and deep, the steam has nowhere to go but up, taking the batter with it. Popover procedures generally call for an oven that's at 450° to 500°F. Since the batter is high moisture, steam forms and drives upward before the protein or starch has an opportunity to set.

✦ Heat oven to 450°F and place lightly greased muffin tin or popover pan on a rack in position C.

✦ When the oven and pan are rippin' hot, pour 1 tablespoon of the fat into the bottom of each cup to coat.

✦ Pour enough batter in each hole to come just a little more than halfway up the side and bake for 15 minutes.

Notes:

The first crepe will be for the dog. Trust me. I'm not sure if it has to do with heat moderation or the amount of fat, but I've never managed to make the first of the batch turn out.

✦ Reduce the oven to 350°F and bake another 10 minutes. The popovers should rise a good 1 to 1½ inches out of the cup and turn golden brown and delicious.

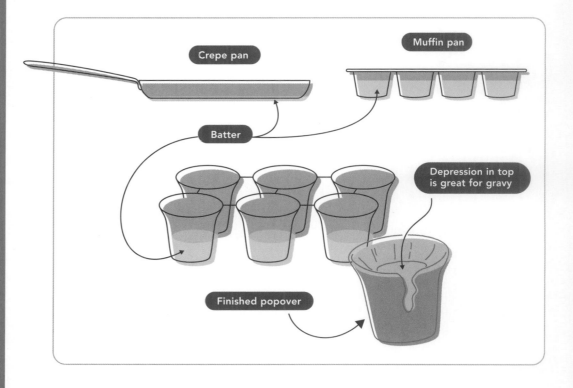

Since there's so much more heat around the perimeter of the batter, the walls are going to rise quickly leaving a hole in the middle.

This is normal. Just consider yourself lucky. Now you have a place to pour your gravy.

Note: What to do with leftover popovers? I have a friend whose name shall remain unspoken who takes the soft shriveled leftover popovers, lines them up in a deep casserole, pours cream and cheese over them and bakes the whole mess at 400°F. The steam from the cream moves into the popovers and pops them again. When it's brown she serves it like a casserole. Sound disgusting? It is...not.

Funnel Cake

THE WET WORKS 1

INGREDIENT	Weight		Volume	Count	Prep
Water	227 g	8 oz	1 cup		
Unsalted butter	85 g	3 oz	6 tablespoons		

THE DRY GOODS

INGREDIENT	Weight		Volume	Count	Prep
Sugar	14 g	$\frac{1}{2}$ oz	1 tablespoon		
Salt			$\frac{1}{8}$ teaspoon		
All-purpose flour	135 g	$4\frac{3}{4}$ oz	1 cup		

THE WET WORKS 2

INGREDIENT	Weight		Volume	Count	Prep
Eggs	200 g	7 oz		4 large	
Egg whites	60 g	2 oz		2 large	

THE EXTRAS

Vegetable oil, for frying ✦ Powdered sugar, for topping

✦ Heat about 1½ inches of oil in a pan to 375°F.

✦ In a heavy-bottomed saucepan, combine the water, butter, sugar, and salt and bring to a boil.

✦ Add the flour and stir with a wooden spoon or rubber spatula until it is completely incorporated and the dough forms a ball. Transfer the dough into the bowl of an electric stand mixer fitted with the paddle attachment and let cool for 3 to 4 minutes.

✦ With the mixer on stir or its lowest speed, add the eggs and the egg whites, 1 at a time, making sure the first egg is completely incorporated before continuing. Once all eggs have been added and the mixture is smooth put the dough into a plastic freezer bag with one lower corner snipped off, or a piping bag fitted with a #12 tip.

✦ Pipe the dough into the hot oil, making a free-form lattice pattern and cook until brown, flipping once using a pancake turner. Remove cake from oil, drain on a cooling rack placed over paper towels or newspaper, and top with powdered sugar. Continue until all of the batter is used.

Yield: 10 cakes

Hardware:

3 small glass bowls for separating eggs

Large, deep frying pan or cast-iron skillet

Medium-sized heavy-bottomed saucepan

Wooden spoon or rubber spatula

Stand mixer fitted with paddle attachment

Plastic freezer bag or piping bag fitted with #12 pastry tip

Pancake turner

Cooling rack

Paper towels or sheets of newspaper

Notes:

What happens when Pâte à Choux meets the county fair.

Hardware:

Digital scale

3 bowls for separating eggs

Large metal bowl

Stand mixer, electric hand mixer, or balloon whisk

Dry measuring cups

Wet measuring cups

Medium saucepan

Wooden spoon

Pastry bag fitted with round tip or plastic bag with one corner snipped off

Parchment paper

2 half sheet pans or jellyroll pans

Paring knife

Notes:

It's also called choux paste, choux pastry, or just cream-puff pastry.

Pâte à Choux

The literal translation from the French of pâte à choux is "paste of cabbage." I guess if you tried hard you might make something that looks like...actually I can't imagine anything about this stuff that looks like cabbage—those wacky French! Pâte à choux is made like no other type of pastry dough. The dough is sticky and, yes, paste-like, and the addition of eggs makes for large, irregular puffs. This is a pastry that's made for filling—you can poke a hole in the side and pipe in pastry cream or split and fill them with ice cream; a savory puff can be filled with chicken salad or any other combination of ingredients.

THE WET WORKS 1

INGREDIENT	Weight		Volume	Count	Prep
Water	227 g	8 oz	1 cup		
Unsalted butter	85 g	3 oz	6 tablespons	³/₄ stick	
Sugar	14 g	¹/₂ oz	1 tablespoon		
Salt			¹/₈ teaspoon		

THE DRY GOODS

INGREDIENT	Weight		Volume	Count	Prep
All-purpose flour	163 g	5³/₄ oz	1¹/₈ cups		

THE WET WORKS 2

INGREDIENT	Weight		Volume	Count	Prep
Eggs	200 g	7 oz	1 cup	about 4 large	
Egg whites	60 g	2 oz		2 large	

✦ ✦ ✦ ✦ ✦

✦ Place an oven rack in position B and preheat to 425°F.

✦ Line 2 half sheet or jellyroll pans with parchment and set aside.

✦ Combine the water, butter, sugar, and salt in a medium saucepan over medium heat and bring to a boil, stirring occasionally.

✦ Add in the flour and remove from heat. Work the flour into the liquid mixture and return to the heat, continuing to work it until all the flour is incorporated and the mixture forms a ball.

✦ Transfer the mixture into the bowl of a stand mixer and let cool for 3 or 4 minutes. With the mixer set for Stir or the lowest speed, add the eggs, 1 at a time, making sure each egg is completely incorporated before continuing.

✦ Once all the eggs have been added and the mixture is smooth, immediately place the dough into a plastic bag with one corner cut off, or a pastry bag fitted with a round tip. Pipe the pâte à choux into golf ball–size shapes onto the parchment-lined sheet pans. Space them 2 inches apart from each other.

✦ Bake for 10 minutes, then turn the oven down to 350°F and bake another 10 minutes, or until golden brown.

✦ Remove the pâte à choux from the oven and immediately pierce them with a paring knife to release steam.

✦ Unfilled, these will keep air-tight in a zip-top bag for 3 days, or freeze for up to two months.

Yield: 4 dozen bite-size cream puffs

✦ ✦ ✦ ✦ ✦

Variation

To make savory pâte à choux, omit the sugar and increase the amount of salt to 1 teaspoon.

Notes:

As always, so many to thank...

Up in New York

First and foremost to my editor at STC, Marisa Bulzone, whose patience, style, wisdom, vision, calm, taste, and skill know no bounds, though I do think I pushed her on this third one a couple of times. STC's design director, Galen Smith, is responsible for the look of this book and I think it's dang spiffy. Thanks Galen...love that green. Kim Tyner is STC's production director, and she always manages to get my books printed and in the stores when they're supposed to be there. I don't know how she does it, but she does.

I'm also grateful to my publicist Amy Voll, who could have traded me in for a handful of less troublesome clients long ago, but didn't. At least, not as of this printing.

Up in Vermont

The King Arthur Flour Company donated a whole lot of flour, which was used for the testing of most of the recipes in this book—and for that I am very grateful.

Down in Atlanta

Vanessa Parker runs the production kitchen on a certain cooking show that I make and she helped develop just about every recipe in this book. She kept the records and managed the entire project and yet still has all her hair and even manages to smile from time to time. I couldn't have done this without her. Ditto Tamie Cook, my research coordinator and, come to think of it, coordinator in general—who can fix any recipe I can make a mess of and can even tell me what city I'm supposed to be in tomorrow.

Denise Burns, Michael Griffith, and Phyllis Sauls (my mom) served as "civilian" testers and did a darned fine job. Thanks for choking down all those cookies.

And then there's my wife DeAnna, who runs our companies, raises our daughter Zoey (all but single-handedly), and actually manages to keep my planets from spinning off into the chaos of the cosmos at the same time.

Published in 2004 by
Stewart, Tabori & Chang
An imprint of Harry N. Abrams, Inc.

Cataloging-in-Publication Data is on file with the Library of Congress

ISBN-13: 978-1-58479-341-0
ISBN-10: 1-58479-341-4

Edited by Marisa Bulzone
Designed by Galen Smith, Nancy Leonard, Jessi Rymill, and Danny Maloney
Graphic Production by Kim Tyner
Illustrations based on sketches by Alton Brown, rendered by Eric Cole

The text of this book was composed in New Century Schoolbook and Avenir

Printed in China

10 9 8 7 6 5 4 3

HNA
harry n. abrams, inc.
a subsidiary of La Martinière Groupe
115 West 18th Street
New York, NY 10011
www.hnabooks.com